T0074272

Springer Theses

Recognizing Outstanding Ph.D. Research

For further volumes:
http://www.springer.com/series/8790

Aims and Scope

The series "Springer Theses" brings together a selection of the very best Ph.D. theses from around the world and across the physical sciences. Nominated and endorsed by two recognized specialists, each published volume has been selected for its scientific excellence and the high impact of its contents for the pertinent field of research. For greater accessibility to non-specialists, the published versions include an extended introduction, as well as a foreword by the student's supervisor explaining the special relevance of the work for the field. As a whole, the series will provide a valuable resource both for newcomers to the research fields described, and for other scientists seeking detailed background information on special questions. Finally, it provides an accredited documentation of the valuable contributions made by today's younger generation of scientists.

Theses are accepted into the series by invited nomination only and must fulfill all of the following criteria

- They must be written in good English.
- The topic should fall within the confines of Chemistry, Physics and related interdisciplinary fields such as Materials, Nanoscience, Chemical Engineering, Complex Systems and Biophysics.
- The work reported in the thesis must represent a significant scientific advance.
- If the thesis includes previously published material, permission to reproduce this must be gained from the respective copyright holder.
- They must have been examined and passed during the 12 months prior to nomination.
- Each thesis should include a foreword by the supervisor outlining the significance of its content.
- The theses should have a clearly defined structure including an introduction accessible to scientists and experts in that particular field.

Christina A. Knapek

Phase Transitions in Two-Dimensional Complex Plasmas

Doctoral Thesis accepted by
Ludwig-Maximilians-Universität, München, Germany

 Springer

Author
Dr. Christina A. Knapek
Max Planck Institute for Extraterrestrial
   Physics
Giessenbachstrasse
85740 Garching
Germany
e-mail: knapek@mpe.mpg.de

Supervisor
Prof. Dr. Gregor E. Morfill
Max Planck Institute for Extraterrestrial
   Physics
Giessenbachstrasse
85740 Garching
Germany
e-mail: gem@mpe.mpg.de

ISSN 2190-5053
ISBN 978-3-642-19670-6
DOI 10.1007/978-3-642-19671-3
Springer Heidelberg Dordrecht London New York

e-ISSN 2190-5061
e-ISBN 978-3-642-19671-3

Cover design: eStudio Calamar, Berlin/Figueres

Printed on acid-free paper

Springer is part of Springer Science+Business Media (www.springer.com)

Supervisor's Foreword

Two-dimensional particle systems have been widely investigated during the last decades. Prominent examples are the 2D crystals of electrons on the surface of liquid helium and the research in the field of 2D colloids—nanometer-sized particles suspended in a liquid and restrained to a single plane. Since the Nobel Prize for physics in 2010 was awarded for groundbreaking studies of Graphene, a single layer of carbon, two dimensional materials gained even more attention, both in science and also in public.

The experimental studies presented in this thesis refer to 2D systems belonging to the field of "complex plasmas".

A complex plasma, in the most basic sense, consists of a mixture of electrons, ions, neutral gas atoms and charged micrometer-sized particles. These complex plasmas can spontaneously self-organize and form "plasma crystals". The plasma crystals were predicted theoretically by Ikezi in 1986, and were experimentally discovered in 1994 by Thomas et al. In the laboratory, complex plasmas can nowadays be easily generated by inserting, e.g., plastic spheres into plasma reactors, achieving many-particle systems in a multitude of states, which serve as an excellent model system for studying the behavior of the interaction of charged particles in strongly coupled systems. Some of the enormous advantages of complex plasmas are that they are optically thin, observable even with low magnification by conventional video cameras, and they are only weakly damped by neutral gas friction (in contrast to the overdamped colloidal matter), resulting in typical time scales of dynamical processes in the range of 10–100 Hz.

The particles usually arrange themselves in 3D structures, layers of particles stacked above each other. But by careful selection of experimental parameters, it is possible to generate a single layer of particles, yielding a unique opportunity to observe 2D systems of interacting particles at the kinetic level by direct imaging of the individual particle motion within that layer.

The work presented by Dr. Knapek deals with the phase state and phase transitions in the 2D complex plasma, leading to the question how significantly the thermodynamical behavior in two dimensions differs from that in three dimensions.

One important aspect, following from the Mermin-Wagner theorem (1966) and already stated by Peierls in 1935, is the absence of true long-range order in 1D and 2D due to the loss of long-range correlations. Without order, there can be no phase transition in the thermodynamical sense. Fortunately, a new type of a so-called "topological phase transition" to a state of quasi long-range order in two dimensions was soon introduced and the theoretical model was first established by Beresinksii in 1971, and further developed by Kosterlitz and Thouless, leading to the well-known Berezinskii-Kosterlitz-Thouless transition. The model explains a mechanism mediated by the condensation of topological defects (or vortices), and has been applied, e.g., to the XY spin model, 2D neutral superfluids, and also to the solid–liquid phase transition in 2D crystals, where the role of vortices is adopted by lattice defects. Other models attempting to explain the nature of phase transitions in 2D crystals, e.g. a grain-boundary (chain-like structures of concatenated lattice defects) induced melting or the density wave theory, exist, but did not have similar impact.

A different approach to the description of the thermodynamical behavior of a 2D system can be obtained by extending the kinetic theory developed by Yakov Frenkel and published in 1945. Frenkel's theory connects the number of lattice defects to the thermodynamical quantity temperature by means of free energy calculations and considerations of domain formation in a (3D) molecule crystal. Based on this concept, a modified theory was developed and applied to the 2D case to explain the experimental results.

This thesis contains the experimental results of two carefully conducted experiments with single layer complex plasmas: the first one demonstrating a new straightforward method to determine the coupling strength—one of the fundamental quantities characterizing the phase state in the many-particle system and the second one concerning the non-equilibrium phase transition occurring during a rapid cooling process. An extensive analysis of global and local structural order parameters as well as the particle energy as time-dependent quantities during the process of recrystallization from an unordered to an ordered state provides a deep insight into the underlying mechanisms of 2D phase transitions, much better than anything available so far. By connecting the local order—determined by lattice defects—to the dynamical properties of the particle ensemble, a scale-free behavior during the transition was discovered, implying universality. This remarkable new finding, which is not compatible with the standard equilibrium physics relationship has triggered a great deal of discussion among the experts. Thus the analysis of the dynamic evolution of lattice defects in complex plasmas has enabled a new understanding of the fundamental stability principles of condensed matter and the self-organization within the ensemble. If the findings could be identified as generic properties, this could be of high interest in surface and membrane physics and also in nanoengineering. The possible interpretation based on Frenkel's work (described above) provides a new approach towards understanding dynamical self-organization processes, which will trigger much further research.

Garching, April 2011   Prof. Dr. Dr. h.c. Gregor E. Morfill

Acknowledgments

My thanks go to Prof. Dr. Gregor Morfill, who gave me the choice for a thesis topic between something with industrial applications of complex plasmas and something concerning basic science, namely phase transitions. If you read the thesis, you know what I choose. But seriously, this is a very interesting field, and I want to thank Prof. Morfill for the opportunity to study it, and for helpful discussions on the theoretical part.

The most important help I got from Dr. Dmitry Samsonov, whose experimental setup I used for my experiments, and who always had time for discussion and lots of patience to explain things, be it basics of electronics or matters of physics and data analysis. Thank you, Dmitry!

Also many thanks to Dr. Sergey Zhdanov for his help and discussions on theoretical aspects of phase transitions and data analysis, and most important, for the simulations of 2D recrystallization.

I thank Dr. Alexey Ivlev for the theoretical part of the estimation of the coupling parameter, and for the general help in case of questions on theory. Thanks also to Dr. Boris Klumov for preparing simulations for the paper on the same topic.

Special thanks go to Dr. Uwe Konopka. First, for always having time to answer any question. Second, for practically always actually knowing an answer to any question. Third, for having patience and for giving me the opportunity to work in PlasmaLab.

Further I want to thank our secretary Angelika Langer for her support in any administrative problem, the steady supply with coffee, and most important for moral support and listening and all the good advice.

Next I want to thank Robert Sütterlin for very helpful discussions on error analysis and pixel noise estimations, and for any help regarding computer problems.

I also want to specially thank Dr. Milenko Rubin-Zuzic who made a talk on complex plasmas years ago for a group of students, of whom I was part of, which I found so interesting that I joined the group. I also thank Dr. Richard Quinn, who was the supervisor for my diploma thesis in the Complex Plasma group.

Thanks also to Dr. Slobodan Mitić, for providing company at outside coffee breaks, and inventing such funny distracting activities like office golf.

I don't know where to begin with the rest, because everyone in our group is always ready to answer any question and to help with problems. So I can only thank all other people of the Complex Plasma group, for all the help and great atmosphere.

Thanks to Ralf Heidemann, Dr. Manis Chauduri (thanks for Indian hot food), Peter Huber, Dr. Mierk Schwabe (thanks for parabolic flight support), Dr. Michael Kretschmer (also thanks for parabolic flights), Martin Fink, Dr. Tetjana Antonova, Dr. Sergey Khrapak, Dr. Vladimir Nosenko, Dr. Mikhail Pustylnik, Lisa Wörner, Dr. Victoria Yaroshenko, Dr. Herwig Höfner, Dr. Julia Zimmermann, Dr. Lenaic Couedel, Dr. Tetsuji Shimizu, Dr. Hubertus Thomas, Prof. Dr. Markus Thoma, Prof. Dr. Vadim Tsytovich, and Elsbeth Collmar.

Not to forget the people in the other building (especially thanks to you for providing room for after work barbecues and such): Philip Brandt, Chengran Du, Dr. Pintu Bandyopadhyay, Ke Jiang, Dr. Yang-fang Li, and Dr. Satoshi Shimizu.

And a big thank you goes to the engineers, without whom nothing would work: Tanja Hagl, Karl Tarantik, Günter Wildgruber, Günther Stadler, Dr. Hermann Rothermel, Christian Rau, Sebastian Albrecht, and Valeriy Yaroshenko.

I apologize to anyone I forgot to mention unintentionally. Thank you to those, too!

Finally, my thanks go to my family: my sister Petra for barbecue evenings, recent football afternoons, and talking about things other than physics, and to my parents Renate and Erwin for their support and trust in me. Thank you for causing me to be here in the first place, financing my time at the university, and all advice you gave me throughout my life.

Really finally now, I thank my boyfriend Daniel Mohr. When I asked him how to refer to him in the acknowledgements, he insisted on the following formulation of the acknowledgement: "I thank Daniel Mohr for private communication and cooking." I allow myself to add to this, that I also thank you, Daniel, for your support in any situation and for being a most important part of my life!

Contents

Chapter 1
Introduction

The mechanisms of phase transitions in two-dimensional systems are subject to extensive investigation, e.g. in monolayer crystals of molecules at interfaces [1], in trapped atomic gases [2], or 2D superconducting vortex lattices [3]. Several theoretical studies describe the possible nature of such phase transitions [4–8], often affirmed by results of computer simulations [9–11]. A variety of experiments have been performed with different two-dimensional or quasi-2D systems to confirm the theoretical predictions. Works include the two-dimensional electron solid [12, 13], or X-ray scattering studies on the freezing of monolayer structures of xenon on graphite [14, 15].

Most prominent in this field of research are colloidal systems of particles immersed in a fluid, because they provide the possibility to generate two-dimensional, easily observable particle systems, similar to the complex plasma [16–19]. Colloid systems are usually strongly overdamped, which leads to the necessity of long observation times, but—as complex plasmas—they provide the interesting feature of direct optical observation of particles trajectories. This makes it possible to characterize the nature of a phase transitions with regard to the dynamical features of the system, and not only by thermodynamical quantities of the ensemble.

This thesis aims to investigate experimentally the dynamical and kinetic properties of two-dimensional complex plasmas with regard to their phase state.

Complex plasmas in the laboratory consist of micron-sized plastic spheres levitated in the sheath of a noble gas discharge. The grains get charged negatively due to fluxes of charged plasma constituents to their surface, and interact via a screened Coulomb potential [20]. They can be made visible to cameras or the eye by illumination with a laser. The particle motion can then be studied directly by tracking the particle trajectories from recorded images, yielding a direct measurement of the dynamical properties, namely the particle coordinates and velocities.

If the experimental conditions are chosen accordingly, the grains can form regular structures, the plasma crystal, with interparticle distances of several $100\,\mu$m, and even real two-dimensional systems can be generated, consisting of a single plane of particles [21–23].

C. A. Knapek, *Phase Transitions in Two-Dimensional Complex Plasmas*,
Springer Theses, DOI: 10.1007/978-3-642-19671-3_1,
© Springer-Verlag Berlin Heidelberg 2011

Methods of phase manipulation include the variation of the neutral gas pressure in the plasma chamber [24, 25], local heating of individual particle by lasers [26, 27], or the application of strong electric pulses to melt an initially crystalline state [28], as it was also done here.

In the work described in this thesis, two-dimensional, initially crystalline, complex plasmas were generated in an asymmetric radio-frequency Argon discharge at low neutral gas pressure. They were studied in their initial state, and melted by negative electric pulses to induce a phase transition and observe the process of recrystallization. The horizontal particle layer was imaged from the top view by a high speed digital camera to ensure high temporal resolution.

In the first step of the data analysis, the uncertainties arising from the pixel-noise of the images and from the uncertainty of the tracking procedure, impose significant restrictions on the spatial resolution of coordinates and consequently on the temporal resolution of the velocities [29, 30]. In the scope of this thesis, an extensive investigation of those uncertainties was performed by simulating different error sources and examining their effects on the further data analysis. This included not only a static analysis of single artificial images, but also the simulation of particle motion and the resulting errors in dynamical properties such as velocities. Further, many different particle-image sizes and levels of pixel-noise were taken into account, yielding a total of 2380 artificial data sets available for statistical interpretation of the errors. The implications of this study are taken into account during the scientific interpretation of all presented experiments.

The first experiment concentrates on the coupling parameter Γ of two-dimensional complex plasmas. For a system of interacting particles, Γ is defined as the ratio of their mean potential energy to their mean kinetic energy. Together with the screening parameter κ, given as the ratio of mean interparticle distance to the Debye length of the charged particle in the plasma, it describes the phase state of the particle system with regard to the coupling strength between individual particles, and it can be used to determine its phase and as a criterion for the occurrence of phase transitions [31, 32]. Phase diagrams (Γ, κ) for 3D complex plasma systems have been obtained numerically [31, 33, 34], but to measure Γ experimentally, the particle charge and temperature, and the screening parameter have to be determined in order to calculate the potential and kinetic energies. The procedures to measure those usually involve additional experimental effort for the charge and κ estimation (e.g. investigation of particle collisions [35], or wave spectra analysis [36]), and a high temporal and spatial resolution for the temperature, and are subject to large uncertainties. Here a method will be introduced which allows to obtain Γ from solely the particle dynamics, i.e. the spatial particle coordinates. The method is applied to a two-dimensional plasma crystal, recorded at a high spatial and temporal resolution. Additionally, the particle charge, screening parameter and particle temperature are measured by conventional methods, and the outcomes for the coupling parameter are compared. It will be shown that the proposed methods gives reliable results consistent with conventional methods, and also follows from simple geometrical approaches.

The second experiment constitutes the main part of this work. The aim was to investigate the phase transition of a two-dimensional complex plasma from an unordered state to an ordered state. Different states of order, and also transitions between them, have been observed in several experiments with quasi-2D and 2D complex plasmas [24–26], and the results have been compared to known theories of two-dimensional melting (reviewed in [37]). Here for the first time the connection between the dynamical and kinetic properties during the crystallization process will be investigated on a fundamental level. From this, the character of the phase transition can be deduced, i.e. a connection between the individual particle motion and the thermodynamical state of the system is established.

As in the first experiment, a two-dimensional plasma crystal was generated in a low pressure, radio-frequency Argon discharge, and then melted by application of a 0.2 s negative electric pulse to wires attached inside the plasma chamber. The short pulse melts the crystal completely, followed by a phase of rapid cooling and recrystallization, during which no external forces were applied to the system, and the damping of particle motion is determined by the friction with the neutral gas atoms. The whole process of melting and crystallization was observed from the top view at a high temporal resolution with 500 recorded frames per second. The particle motion was tracked and velocities were calculated by tracing the particles across consecutive frames.

The velocities yield the time dependent mean kinetic energy of the particles as a thermodynamical quantity, defining the development of the thermodynamical state during recrystallization.

The development of the order in the system was studied with regard to translational and orientational correlation functions [5, 38, 39], calculated at each available time step, yielding parameters such as correlation lengths as a measure for the spatial range of long range order. Additionally, the "local" order at the individual particle positions, affected by the degree of regularity in the crystallizing lattice, and by local distortions of the lattice structure, has been investigated.

All structural parameters clearly indicate a highly disordered state short after the pulse was applied, followed by a rapid change towards the initial, crystalline state. The kinetic particle energy as the thermodynamical property of the system was then connected to the structural properties such as defect fractions, correlation lengths and the local orientational order of nearest neighbor bonds which gives an estimate for the goodness of the lattice structure. It was found that all quantities exhibit a power law dependence on the energy, indicating a scale free transition from a disordered to an ordered state. The reason of the observed behavior is identified as the forming of domains of different lattice orientations, separated by strings of defects, or grain boundaries. During the rapid cooling, the domains grew continuously in size, while the defect fraction decreased. It will be shown that the grain boundaries seem to contain a considerable fraction of the total energy in the particle system, and therefore strongly influence the thermodynamical behavior, leading to the scale free, continuous transition to higher order. A possible theoretical model based on the work of Frenkel [40], consistent with the experimental findings, is presented under reserve of certain assumptions which still have to be confirmed.

The thesis is organized as follows:

Chapter 2 gives a short review of complex plasmas in general, their occurrence in nature and industrial applications, and the basic processes determining the behavior of particles in plasmas.

The experimental setup is described in Chap. 3. There also a short overview of the treatment of measurement uncertainties is given; the complete procedure and all results are compiled in Chap. 9.

Chapter 4 explains the techniques used in the data analysis, including a method for charge estimation and the calculation of structural and statistical properties.

The new method for the estimation of Γ is presented in Chap. 5 with the experimental results and interpretation.

An overview of existing theories of phase transitions in two-dimensional systems is given in Chap. 6, followed by the experimental details and results of the analysis of the recrystallization experiments in Chap. 7.

Chapter 8 summarizes the findings of this thesis, and concludes with an outlook for future studies.

References

1. L. Pauchard, D. Bonn, J. Meunier, Dislocation-mediated melting of a two-dimensional crystal. Nature **384**, 145–147 (1996)
2. Z. Habzibabic, P. Krüger, M. Cheneau, B. Batttelier, J. Dalibard, Beresinkii-Kosterlitz-Thouless crossover in a trapped atomic gas. Nature **441**, 1118–1121 (2006)
3. I. Guillamón, H. Suderow, A. Fernández-Pacheco, J. Sesé, R. Córdoba, J.M. De Teresa, M.R. Ibarra, S. Vieira, Direct observation of melting in a two-dimensional superconducting vortex lattice. Nat. Phys. **5**, 651–655 (2009)
4. J.M. Kosterlitz, D.J. Thouless, Ordering, metastability and phase transitions in two-dimensional systems. J. Phys. C **6**, 1181–1203 (1973)
5. B.I. Halperin, D.R. Nelson, Theory of two-dimensional melting. Phys. Rev. Lett. **41**, 121 (1978)
6. A.P. Young, Melting and the vector Coulomb gas in two dimensions. Phys. Rev. B **19**, 1855 (1979)
7. S.T. Chui, Grain-boundary theory of melting in two dimensions. Phys. Rev. Lett. **48**(14), 933–935 (1982)
8. T.V. Ramakrishnan, M. Yussouff, First-principles order-parameter theory of freezing. Phys. Rev. B **19**(5), 2775–2794 (1979)
9. Y. Saito, Monte Carlo studies of two-dimensional melting: dislocation vector systems. Phys. Rev. B **26**(11), 6239–6253 (1982)
10. Y. Saito, Melting of dislocation vector systems in two dimensions. Phys. Rev. Lett. **48**(16), 1114–1117 (1982)
11. F.F. Abraham, Melting in two dimensions is first order: an isothermal-isobaric Monte Carlo study. Phys. Rev. Lett. **44**(7), 463–466 (1980)
12. C.C. Grimes, G. Adams, Evidence for a liquid-to-crystal phase transition in a classical, two-dimensional sheet of electrons. Phys. Rev. Lett. **42**(12), 795–798 (1979)
13. R.H. Morf, Temperature dependence of the shear modulus and melting of the two-dimensional electron solid. Phys. Rev. Lett. **43**(13), 931–935 (1979)
14. P.A. Heiney, R.J. Birgeneau, G.S. Brown, P.M. Horn, D.E. Moncton, P.W. Stephens, Freezing transition of monolayer xenon on graphite. Phys. Rev. Lett. **48**(2), 104–108 (1982)

15. P.A. Heiney, P.W. Stephens, R.J. Birgeneau, P.M. Horn, D.E. Moncton, X-ray scattering study of the structure and freezing transition of monolayer xenon on graphite. Phys. Rev. B **28**(11), 6416–6434 (1983)

16. C.A. Murray, W.O. Sprenger, R.A. Wenk, Comparison of melting in three and two dimensions: microscopy of colloidal spheres. Phys. Rev. B **42**(1), 688–703 (1990)

17. C.A. Murray, D.H. Van Winkle, Experimental observation of two-stage melting in a classical two-dimensional screened Coulomb system. Phys. Rev. Lett. **58**(12), 1200–1203 (1987)

18. Y. Tang, A.J. Armstrong, R.C. Mockler, W.J. O'sullivan, Free-expansion melting of a colloidal monolayer. Phys. Rev. Lett. **62**(20), 2401–2404 (1989)

19. K. Zahn, R. Lenke, G. Maret, Two-stage melting of paramagnetic colloidal crystals in two dimensions. Phys. Rev. Lett. **82**(13), 2721–2724 (1999)

20. H. Ikezi, Coulomb solid of small particles in plasmas. Phys. Fluids **29**(6), 1764–1766 (1986)

21. H. Thomas, G.E. Morfill, V. Demmel, J. Goree, B. Feuerbacher, D. Möhlmann, Plasma crystal: Coulomb crystallization in a dusty plasma. Phys. Rev. Lett. **73**, 652–655 (1994)

22. J.H. Chu, I. Lin, Direct observation of coulomb crystals and liquids in strongly coupled rf dusty plasmas. Phys. Rev. Lett. **72**, 4009–4012 (1994)

23. S. Hayashi, Y. Tachibana, Observation of Coulomb-crystal formation from carbon particles grown in a methane plasma. Jpn. J. Appl. Phys. **33**, L804–L806 (1994)

24. H.M. Thomas, G.E. Morfill, Melting dynamics of a plasma crystal. Nature **379**, 806–809 (1996)

25. R.A. Quinn, J. Goree, Experimental test of two-dimensional melting through disclination unbinding. Phys. Rev. E **64**, 051404 (2001)

26. V. Nosenko, J. Goree, A. Piel, Laser method of heating monolayer dusty plasmas. Phys. Plasmas **13**, 032106 (2006)

27. M. Wolter, A. Melzer, Laser heating of particles in dusty plasmas. Phys. Rev. E **71**, 036414 (2005)

28. D. Samsonov, S.K. Zhdanov, R.A. Quinn, S.I. Popel, G.E. Morfill, Shock melting of a two-dimensional complex (dusty) plasma. Phys. Rev. Lett. **92**(25), 255004 (2004)

29. Y. Ivanov, A. Melzer, Particle positioning techniques for dusty plasma experiments. Rev. Sci. Instrum. **78**, 033506/1–7 (2007)

30. Y. Feng, J. Goree, B. Liu, Accurate particle position measurement from images. Rev. Sci. Instrum. **78**, 053704/1–10 (2007)

31. S. Hamaguchi, R.T. Farouki, D.H.E. Dubin, Triple point of Yukawa systems. Phys. Rev. E **56**, 4671–4682 (1997)

32. O.S. Vaulina, S.V. Vladimirov, O.F. Petrov, V.E. Fortov, Criteria of phase transitions in a complex plasma. Phys. Rev. Lett. **88** (24), 245002 (2002)

33. E.J. Meijer, D. Frenkel, Melting line of Yukawa system by computer simulation. J. Chem. Phys. **94**(3), 2269–2271 (1990)

34. O. Vaulina, S. Khrapak, G. Morfill, Universal scaling in complex (dusty) plasmas. Phys. Rev. E **66**, 016404 (2002)

35. U. Konopka, L. Ratke, H.M. Thomas, Central collisions of charged dust particles in a plasma. Phys. Rev. Lett. **79**(7), 1269–1272 (1997)

36. S. Nunomura, J. Goree, S. Hu, X. Wang, A. Bhattacharjee, Dispersion relations of longitudinal and transverse waves in two-dimensional screened Coulomb crystals. Phys. Rev. E **65**, 066402/1–11(2002)

37. K.J. Strandburg, Two-dimensional melting. Rev. Mod. Phys. **60**(1), 161–207 (1988)

38. D.C. Wallace, *Statistical Physics of Crystals and Liquids*, chap. 5. (World Scientific Publishing Co. Pte. Ltd., Singapore, 2002)

39. D.R. Nelson, B.I. Halperin, Dislocation-mediated melting in two dimensions. Phys. Rev. B **19**, 2457 (1979)

40. J. Frenkel, *Kinetic Theory of Liquids*. (Dover Publications, Inc., New York, (1955)

Chapter 2
Complex Plasmas

An ionized gas containing ions, electrons and neutral atoms is called a plasma if it meets the conditions quasineutrality and collective behavior. The quasineutrality exists for distances much larger than the Debye length at which the potential caused by a charged particle has dropped to $1/e$ due to the shielding by oppositely charged species. Collective behavior arises when one charged particle interacts with many other charged particles through the Coulomb force. This means much more than one particle has to remain within a Debye sphere. The motion of the constituents of a plasma should further be caused by electromagnetic interaction rather than direct particle collisions, requiring that the frequency of such collisions is much smaller than the plasma frequency—the frequency of typical plasma oscillations [1].

A Complex plasma is a plasma containing micron-sized particles, sometimes referred to as dust particles, as an additional component besides the ions, electrons and neutral gas atoms. These particles get charged inside the plasma due to streams of electrons and ions to their surface, and interact with each other and with the plasma constituents, in turn complicating the behavior of the plasma.

Complex plasmas are present everywhere in nature, from interstellar space and circumstellar clouds [2] to the solar system: as interplanetary dust, in comet tails and planetary rings, and also in the earth magnetosphere and atmosphere [3–6]. The grains can be e.g. ice, silicates, or metallic compounds with a wide variety of shapes. They are subject to radiation pressure and gravity, and, due to their charge, they are affected by the electromagnetic forces of the planetary magnetosphere or solar magnetic winds. Interplanetary dust is responsible for the occurrence of the zodiacal light, while the dust in planetary rings is assumed to be responsible for spoke formation, for example in the Saturn ring system [7–9].

In industrial applications dust dispensed in plasmas can have devastating effects [10]. The chemically active plasmas used for deposition or etching in the manufacturing of microchips are prone to produce unwanted particles of the size of nanometers to micrometers by polymerization and sputtering processes in the ionized gas phase [11]. This contamination of the plasma leads to defects in the fabricated circuits, especially when the grains are of the size of the structures to be generated. In fusion

C. A. Knapek, *Phase Transitions in Two-Dimensional Complex Plasmas*,
Springer Theses, DOI: 10.1007/978-3-642-19671-3_2,
© Springer-Verlag Berlin Heidelberg 2011

reactors, particles are generated in plasma-surface interactions, leading to problems with plasma stability and safety of the operation of the device [12]. On the other hand, applications such as surface processing make use of the grains dispersed in the plasma [10]. Examples are the growth of carbon-based nanostructures on surfaces used for electronic devices such as sensors or data storage technologies [13]. Silicon-based films as used in flat panel displays or solar cells show an improved performance if nanoparticles, generated in the plasma by chemical reactions, are embedded into the film [14]. Other applications include the synthesis of composite materials or coating of particles by plasma processing [15].

To understand the behavior of the particle component, complex plasmas are generated and studied in laboratories on earth, and under microgravity conditions in parabolic flights and on the international space station (ISS) [16–18]. The particles with sizes of usually several μm are inserted into radio-frequency (RF) [19–22] or DC discharges [23, 24] of inert gases in plasma reactors, and can be made observable to the eye and to video cameras by scattering of visible laser light. External parameters such as the reactor geometry, gas pressure, discharge conditions and particle composition and size determine the particle charge and temperature and therefore the dynamical behavior and the state of order of the system. Typical parameters for the discharge are neutral gas pressures of 1 to 100 Pa and ion and electron densities of 10^8–10^9 cm^{-3} in the plasma. The electron temperature lies usually within 1–7 eV, while the ions and the neutral gas atoms are basically at or close to room temperature 0.025–0.03 eV. The dust particles are typically dielectric plastic spheres, with radii in the range of 0.5–30 μ, and they can reach charges of 10^3–10^5 elementary charges [10].

The time scales for grain charging are very short, and an equilibrium state can be reached within fractions of seconds. Compared to e.g. strongly overdamped colloidal systems (particles immersed in a fluid) with equilibration times of days to weeks [25, 26], the lower friction on particles in a plasma environment leads to much shorter equilibration and observation times. Further, complex plasmas are optically thin, allowing the observation of many particle layers. The multitude of possible different states—from gaseous over liquid to solid-like states, in two or three dimensions—and the usual short time scales of physical processes and good conditions for optical data acquisition makes the complex plasma a perfect model system to study particle interactions. A wide range of other phenomena such as dust-acoustic and dust-lattice waves [27–30], and instabilities have been widely investigated (see reviews in [16, 31]). Recently, nonlinear phenomena like solitons [32–35], shock waves [36] and Mach cones [37] were discovered. Also of interest are kinetic studies of fluid dynamics with complex plasmas, e.g. superdiffusion [38] or the recently found electrorheological plasmas [39]. Phase transitions between solid and liquid-like states have been observed in three- and also two-dimensional complex plasmas [40–46].

Purely two-dimensional systems are in general difficult to find in nature. Complex plasmas provide a relatively easy way to generate such a particle system. If the experimental parameters are chosen accordingly, a single horizontal layer of particles can be levitated at an equilibrium height above a horizontal electrode inside a rf discharge chamber. Vertical forces can be neglected in the analysis, provided vertical fluctuations (e.g. caused by charge fluctuations) are small. The particle dynamics are then restricted to a 2D plane, simplifying especially the structural analysis of the system. Also, an instantaneous study of the mutual particle interaction and the resulting dynamics of the system is possible: all neighboring particles can be imaged at once on a 2D imaging device, while in three-dimensional complex plasmas only a cross section of the whole system can be recorded at one time.

The following Sect. 2.1 will in short address the charging process and the particle interaction potential. The different forces acting on the particles dispersed in a plasma, especially in the (earth bound) laboratory, are discussed in Sect. 2.2. Lastly, the fundamental characterization of complex plasmas as coupled systems is given in Sect. 2.3.

2.1 Particle Charge and Interaction

Several processes contribute to the charging of a particle in a plasma. Most important is the collection of charge carriers on the grain surface due to the electron and ion fluxes any surface in contact with a plasma is subject to. The temporal charge evolution is then given by

$$\frac{dQ_D}{dt} = I_i - I_e \tag{2.1}$$

Expressions for the fluxes $I_{e,i}$ can be derived from the OML (orbit motion limited) approximation of isolated grains in a collisionless plasma [47–49]. Also no barrier in the effective potential of the grain is assumed, so that ions can neither be trapped in orbits around the particle nor be reflected by the barrier.

Then [16, 50]

$$I_e = \sqrt{8\pi T_e/m_e} a^2 n_e \exp(e\Phi_f/T_e)$$
$$I_i = \sqrt{8\pi T_i/m_i} a^2 n_i (1 - e\Phi_f/T_i)$$

with the temperatures $T_{e,i}$, masses $m_{e,i}$ and densities $n_{e,i}$ for electrons and ions, respectively, and the particle surface potential Φ_f. The net particle charge can be obtained by solving the equilibrium condition $I_e = I_i$ for Φ_f, which is usually negative with respect to the plasma due to the higher mobility of the electrons. Then $Q_D = C\Phi_f$ with the capacitance C. In case of spherical particles with radius a, the capacitance would be $C = 4\pi\epsilon_0 a$.

It was found that this approximation works well for sufficiently small grains with $a/\lambda_D < 0.2$ with the grain radius a [51, 52]. There also exist models for weakly or strongly collisional regimes and for anisotropic plasmas, which often lead to much smaller charges than the predictions of the OML theory [16, 53, 54].

Another mechanism contributing to the charging is the electron emission. In low temperature plasmas the thermionic emission of electrons can lead to positive particle charges [16]. For complex (dusty) plasmas in space the photoelectric emission of electrons due to fluxes of photons to the particle surface plays an important role [55], as does the secondary electron emission caused by incident primary electrons with energies above a material dependent threshold [56]. In the plasma regime of a typical RF-discharge, as used for the experiments presented later, the contribution to the charge by the emission processes can be neglected.

The charged particle has a linearized Debye length defined by the electron and ion Debye lengths $\lambda_{e,i}$ [51, 57, 58]

$$\frac{1}{\lambda_D^2} = \frac{1}{\lambda_e^2} + \frac{1}{\lambda_i^2} = \frac{e^2 n}{\epsilon_0} \left(\frac{1}{k_B T_e} + \frac{1}{k_B T_i} \right) \tag{2.2}$$

where n is the undisturbed electron and ion density and $T_{e,i}$ are the temperature of electrons and ions, respectively. Inside the plasma sheath of a discharge, where the dust particles are usually located, the ion energy is large due to the Bohm sheath criterion and $\lambda_D \approx \lambda_e$ [59].

The charging time is inversely proportional to the particle size and the plasma density n, $\tau \propto 1/(an)$, and lies in the range of 10^{-6} s. The charging process is stochastic, therefore charge fluctuations appear on a scale defined by the width of the Gaussian charge distribution, which is inversely proportional to $\sqrt{Q_D}$, and fluctuations appear at frequencies below $0.024/\tau$ [50, 60].

The charged particles interact with each other via an electrostatic potential which can be approximated by a screened Coulomb potential (Yukawa-like potential) [61]:

$$\Phi(r) = \frac{Q_D}{4\pi \epsilon_0 r} e^{-\Delta/\lambda_D} \tag{2.3}$$

Here Δ is the average distance between two particles, and for like particles, $\Phi(r)$ is always repulsive. The approximation does not take into account the charge fluctuations, and it assumes a low enough particle density with "isolated" particles.

There exist effective attractive forces between the particles, e.g. the ion shadowing force, which appears for particles in the plasma at distances less than the ion mean free path. Then surrounding particles shield the fluxes of plasma constituents to a particle surface. A similar shadowing force due to neutral gas atoms exists, which can be both attractive or repulsive, depending on the ratio of the temperature of the particle surface to that of the neutral gas. The shadowing forces scale with $1/r$ and therefore gain importance compared to the electrostatic interaction $\propto 1/r^2$ at larger r [16, 62, 63]. Another attractive force is caused by ions streaming past the particles and producing a positive wake behind the negative particle which attracts other negative particles [62, 64]. For the investigation of short-range particle interactions, the Coulomb approximation work well and attractive forces are often neglected.

2.2 Forces

Besides the interparticle forces, several external forces play a major role in the behavior of the particle component in a complex plasma. These forces are shortly introduced in the following section (see e.g. [65] for a detailed compilation).

Electrostatic Force An external electric field $\mathbf{E}$ causes a force which can be well approximated by the electrostatic force $F_{es} = Q_D\mathbf{E} \propto a$ which scales linear with the particle radius a. The Debye sheath surrounding the particles can be neglected as long as the dust particle radius is small compared with the shielding length and the sheath is not distorted [66]. If there is a distortion in the sheath, an additional polarization force $F_P = -Q_D^2\mathbf{E}/(8\pi\epsilon_0\lambda_D)$ has to be added [57]. The electrostatic force is an important tool for particle confinement inside plasma chambers, or for complex plasma manipulation.

Ion Drag Force Ions exchange momentum with dust grains due to Coulomb interaction ("orbit force", o) and collection of ions ("collection force", c), the later leading to a momentum transfer. The result force $F_{c,o} = n_i m_i \sigma_m^{c,\bar{o}} v\mathbf{v}$ depends on the ion velocity $\mathbf{v}$ and the cross-sections $\sigma_m^{c,o}$ for each process. Both parts of the ion drag force scale with a^2, due to the quadratic dependence of the cross-section on the particle radius. Especially in the plasma sheath region between the plasma and a wall, ions are accelerated towards the wall due to the electric field inside the sheath which points to the wall. The ion drag force therefore tends to push particles located in the plasma sheath toward the walls [67–70].

Neutral Drag Force Another force exerted on the particles is the neutral gas drag due to the momentum transfer in collisions of particles with neutral gas atoms. For small relative velocities between gas atoms and dust particles this force can be approximated by the Epstein drag force as $F_{nd} = -4/3\delta a^2 m_N n_N v_{th,N}(\mathbf{u}_D - \mathbf{u}_N)$, with the gas mass and density m_N, n_N and the thermal velocity of the gas $v_{th,N}$. $\mathbf{u}_{D,N}$ are the mean velocities of the particles and neutral gas atoms, respectively [71]. The coefficient δ has to be chosen according to the mechanism of the reflection of gas atoms from the particle surface. Epstein calculated values for spheres for specular ($\delta = 1$) and diffuse ($\delta = 1 + 9\pi/64 = 1.442$) reflection, the latter being the case for a thermal non-conducting material. Recent measurements using complex plasmas resulted in similar values in the range 1.26–1.48 [59, 72].

Thermophoretic Force In case of a temperature gradient in the neutral gas component, a force called thermophoretic force acts on the particles. It is caused by the higher momentum transfer on the side of the higher temperature and points to regions with lower temperature. Therefore it is proportional to the thermal conductivity of the gas κ_T, the temperature gradient and the squared particle radius: $F_{th} \propto a^2\kappa_T\nabla T_N$. Thermophoresis can be used to levitate particles in a plasma chamber against gravity into regions in the bulk plasma, providing similar conditions as experiments in microgravity, and it can lead to other interesting phenomena such as convection, circulations and formation of structures [73–76].

Gravitational Force On earth, the particles are subject to gravity. The gravitational force $F_g = m_D g = 4/3\pi a^3 n\rho_D g$ for particles with the mass density ρ_D scales with a^3

and becomes especially important for large particles with radii $> 10 \, \mu m$, or particles with a high mass density. To levitate them inside a plasma, an opposing force has to be present. In a typical parallel plate RF discharge, the electric field $\mathbf{E}(z)$ in the plasma sheath regions between electrodes and bulk plasma depends linear on the distance z to the surface. For horizontal electrodes, it points upwards from the lower electrode and can counteract gravity and levitate the particles at the height ζ where $m_D g = Q_D \mathbf{E}(\zeta)$.

In summary, the forces acting on a particle are listed in their order of magnitude as it is valid for small particles ($a < 1 \, \mu m$), including their dependence on the particle radius:

$$F = F_{es} + F_{id} + F_{nd} + F_{th} + F_g$$
$$F_{es} \propto a; \quad F_{id}, F_{nd}, F_{th} \propto a^2; \quad F_g \propto a^3 \tag{2.4}$$

The electrostatic force is dominant, followed by the orbit-force part of the ion drag force. The neutral drag force and thermophoresis are usually of the same magnitude. As the particle radius becomes larger, all forces gain in magnitude, but especially the ion drag and gravitational forces will exceed F_{es} for particles with $a > 10 \, \mu m$.

2.3 Characterization of a Complex Plasma

Two dimensionless quantities, namely the coupling strength Γ and the screening parameter κ, are used to characterize a complex plasma [77].

The screening parameter κ is the ratio of the average interparticle distance Δ to the screening length λ_D of the dust grains:

$$\kappa = \Delta / \lambda_D \tag{2.5}$$

The factor $e^{-\kappa}$ appeared in (2.3) and reduces the potential strength of the pure Coulomb interaction potential. When the screening length is small compared with the interparticle distance, the neighboring particles will see a smaller effective charge, and the potential will be diminished.

The coupling strength Γ is the ratio of the mean potential energy to the mean kinetic energy of the particles. The potential energy is defined by the average position a particle occupies in the potential well created by its charged neighboring particles. Using the Yukawa-type interaction potential Φ given in the former paragraph, the potential energy can be calculated from $E_{\text{pot}} = Q\Phi$. The kinetic energy E_{kin} is derived from the average thermal motion of the particles with a temperature T, yielding $\frac{1}{2} k_B T$ per degree of freedom with the Boltzmann constant k_B. Then

$$\Gamma = \frac{\langle E_{\text{pot}} \rangle}{\langle E_{\text{kin}} \rangle} = \frac{Q^2}{4\pi\epsilon_0 \Delta \cdot 0.5 n_{\text{dof}} k_B T} F(\kappa) \tag{2.6}$$

with the number of degrees of freedom $n_{\text{dof}} = 2$ (2D) or $n_{\text{dof}} = 3$ (3D).

The factor $F(\kappa)$ describes the influence of the screening on the mean potential energy and depends on the underlying theory used to describe the particle system. In the one-component-plasma (OCP) limit with a screened Coulomb interaction potential, $F(\kappa)$ would simply be $\exp(-\kappa)$ [78].

If the average kinetic energy of particles in a complex plasma exceeds their average potential energy, the particles will not be caught in the potential well created by surrounding particles, and the system will be in a liquid-like or gaseous state. In the opposite case of strong coupling, for $E_{kin} \ll E_{pot}$, a crystalline structure of dust grains, the plasma crystal, can form. The existence of such a structure, also called Coulomb crystal, was predicted theoretically by Ikezi in 1986 [78], and later verified by the first observations of plasma crystals in gas discharges [19–21]. Phase transitions to or from this solid state of complex plasmas were investigated in simulations [41] even prior to the first experimental discovery of the crystalline phase. The good experimental accessibility of complex plasmas, mentioned in the introduction of this chapter, offers a opportunity for the study of such phase transitions. The knowledge of Γ in that case is of high importance, since it contains valuable information on the thermodynamic state of the system, and can be used to characterize the transitions.

For 3D particle systems interacting by a Yukawa-type potential, (Γ, κ) phase diagrams were calculated in molecular-dynamics simulations for the OCP limit, indicating melting lines between a fluid phase and two solid phases with either a fcc (face-centered-cubic) or bcc (body-centered-cubic) lattice [79–81]. Later the phase diagrams were revised by taking into account the lattice dynamics, which modifies $F(\kappa)$ [77, 82–84]. A method to obtain Γ from easily accessible data in experiments with 2D complex plasmas will be presented later in this thesis in Chap. 5, where also an expression for $F(\kappa)$, specifically for two-dimensional systems, will be given.

References

1. F.F. Chen, *Introduction to Plasma Physics and Controlled Fusion*. Plasma Physics, vol. 1, Chap. 1.2 (Plenum Press, New York, 1984)
2. U. de Angelis, The physics of dusty plasmas. Phys. Scripta **45**, 465–474 (1992)
3. D.A. Mendis, A postencounter view of comets. Ann. Rev. Astron. Astrophys. **26**, 11–49 (1988)
4. C.K. Goertz, Dusty plasmas in the solar system. Rev. Geophys. **27**(2), 271–292 (1989)
5. O. Havnes, U. de Angelis, R. Bingham, C.K. Goertz, G.E. Morfill, V. Tsytovich, On the role of dust in the summer mesopause. J. Atmos. Terr. Phys. **52**, 637–643 (1990)
6. O. Havnes, T. Aslaksen, A. Brattli, Charged dust in the earth's middle atmosphere. Phys. Scripta T **89**, 133–137 (2001)
7. C.K. Goertz, G. Morfill, A model for the formation of spokes in Saturn's ring. Icarus **53**, 219–229 (1983)
8. E. Grün, G.E. Morfill, R.J. Terrile, T.V. Johnson, G. Schwehm, The evolution of spokes in Saturn's b ring. Icarus **54**, 227–252 (1983)
9. G.E. Morfill, H.M. Thomas, Spoke formation under moving plasma clouds—The Goertz–Morfill model revisited. Icarus **179**, 539–542 (2005)
10. S.V. Vladimirov, K. Ostrikov, A.A. Samarian, Physics and Applications of Complex Plasmas. (Imperial College Press, London, 2005)

11. A. Bouchoule, Technological Impacts of Dusty Plasmas. in *Dusty Plasmas: Physics, Chemistry and Technological Impacts in Plasma Processing*, chap. 4, ed. by A. Bouchoule (Wiley, Chichester, 1999)

12. J. Winter, Dust: a new challenge in nuclear fusion research? Phys. Plasmas **7**(10), 3862–3866 (2000)

13. M. Meyyappan, L. Delzeit, A. Cassell, D. Hash, Carbon nanotube growth by PECVD: a review. Plasma Sour. Sci. Technol. **12**, 205–216 (2003)

14. J. Perrin, J. Schmitt, C. Hollenstein, A. Howling, L. Sansonnens, The physics of plasma-enhanced chemical vapour deposition for large-area coating: industrial application to flat panel displays and solar cells. Plasma Phys. Control Fusion **42**, 353–363 (2000)

15. E. Stoffels, W.W. Stoffels, H. Kersten, G.H.P.M. Swinkels, G.M.W. Kroesen, Surface processes of dust particles in lower pressure plasmas. Phys. Scripta. T **89**, 168–172 (2001)

16. V.E. Fortov, A.V. Ivles, S.A. Khrapak, A.G. Khrapak, G.E. Morfill, Complex (dusty) plasmas: current status, open issues, perspectives. Phys. Rep. **421**, 1–103 (2005)

17. A.P. Nefedov et al., PKE-Nefedov: plasma crystal experiments on the international space station. New. J. Phys. **5**, 33.1–33.10 (2003)

18. H.M. Thomas, G.E. Morfill, V.E. Fortov, A.V. Ivlev, V.I. Molotkov, A.M. Lipaev, T. Hagl, H. Rothermel, S.A. Khrapak, R.K. Suetterlin, M. Rubin-Zuzic, M. Petrov, V.I. Tokarev, S.K. Krikalev, Complex plasma laboratory PK-3 plus on the international space station. New. J. Phys. **10**, 033036 (2008)

19. J.H. Chu, I. Lin, Direct observation of coulomb crystals and liquids in strongly coupled rf dusty plasmas. Phys. Rev. Lett. **72**, 4009–4012 (1994)

20. H. Thomas, G.E. Morfill, V. Demmel, J. Goree, B. Feuerbacher, D. Möhlmann, Plasma crystal: coulomb crystallization in a dusty plasma. Phys. Rev. Lett **73**, 652–655 (1994)

21. Y. Hayashi, S. Tachibana, Observation of Coulomb-crystal formation from carbon particles grown in a methane plasma. Jpn J. Appl. Phys. **33**, L804–L806 (1994)

22. A. Melzer, T. Trottenberg, A. Piel, Experimental determination of the charge on dust particles forming Coulomb lattices. Phys. Lett. A **191**, 301–307 (1994)

23. V.E. Fortov, A.P. Nefedov, V.M. Torchinskii, V.I. Molotkov, A.G. Khrapak, O.F. Petrov, K.F. Volykhin, Crystallization of a dusty plasma in the positive column of a glow discharge. JETP Lett. **64**, 92–98 (1996)

24. S. Mitić, B.A. Klumov, U. Konopka, M.H. Thoma, G.E. Morfill, Structural properties of complex plasmas in a homogenious discharge. Phys. Rev. Lett. **101**, 125002 (2008)

25. C.A. Murray, W.O. Sprenger, R.A. Wenk, Comparison of melting in three- and two-dimensions: microscopy of colloidal spheres. Phys. Rev. B **42**(1), 688–703 (1990)

26. D.G. Grier, C.A. Murray, The microscopic dynamics of freezing in supercooled colloidal fluids. J. Chem. Phys. **100**(12), 9088–9095 (1994)

27. M. Zuzic, H.M. Thomas, G.E. Morfill, Wave propagation and damping in plasma crystals. J. Vac. Sci. Technol. A **14**(2), 496–500 (1995)

28. J.B. Pieper, J. Goree, Dispersion of plasma dust accoustic waves in the strong-coupling regime. Phys. Rev. Lett. **77**(15), 3137–3140 (1996)

29. A. Homann, A. Melzer, S. Peters, R. Madani, A. Piel, Laser-excited dust lattice waves in plasma crystals. Phys. Lett. A **242**, 173–180 (1998)

30. S. Nunomura, J. Goree, S. Hu, X. Wang, A. Bhattacharjee, Dispersion relations of longitudinal and transverse waves in two-dimensional screened Coulomb crystals. Phys. Rev. E **65**, 066402/1–11 (2002)

31. G.E. Morfill, A.V. Ivlev, Complex plasmas: an interdisciplinary research field. Rev. Mod. Phys. **81**, 1353–1404 (2008)

32. D. Samsonov, A.V. Ivlev, R.A. Quinn, G. Morfill, S. Zhdanov, Dissipative longitudinal solitons in a two-dimensional strongly coupled compex (dusty) plasma. Phys. Rev. Lett. **88**(9), 095004 (2002)

33. V. Nosenko, S. Nunomura, J. Goree, Nonlinear compressional pulses in a 2D crystallized dusty plasma. Phys. Rev. Lett. **88**(21), 215002 (2002)

34. S.K. Zhdanov, D. Samsonov, G.E. Morfill, Anisotropic plasma crystal solitons. Phys. Rev. E **66**, 026411 (2002)
35. R. Heidemann, S. Zhdanov, R. Sütterlin, H.M. Thomas, G.E. Morfill, Dissipative dark soliton in a complex plasma. Phys. Rev. Lett. **102**, 135002 (2009)
36. D. Samsonov, S.K. Zhdanov, R.A. Quinn, S.I. Popel, G.E. Morfill, Shock melting of a two-dimensional complex (dusty) plasma. Phys. Rev. Lett. **92**(25), 255004 (2004)
37. D. Samsonov, J. Goree, Z.W. Ma, A. Bhattacharjee, H.M. Thomas, G.E. Morfill, Mach cones in a Coulomb lattice and a dusty plasma. Phys. Rev. Lett. **83**(18), 3649–3652 (1999)
38. S. Ratynskaia, K. Rypdal, C. Knapek, S. Khrapak, A.V. Milovanov, A. Ivlev, J.J. Rasmussen, G.E. Morfill, Superdiffusion and viscoelastic vortex flows in a two-dimensional complex plasma. Phys. Rev. Lett. **96**, 105010 (2006)
39. A.V. Ivlev, G.E. Morfill, H.M. Thomas, C. Räth, G. Joyce, P. Huber, R. Kompaneets, V.E. Fortov, A.M. Lipaev, V.I. Molotkov, T. Reiter, M. Turin, P. Vinogradov, First observation of electrorheological plasmas. Phys. Rev. Lett. **100**, 095003 (2008)
40. H. Thomas, G.E. Morfill, Solid liquid gaseous phase transitions in plasma crystals. J. Vac. Sci. Technol. A **14**(2), 501–505 (1995)
41. R.T. Farouki, S. Hamaguchi, Phase transitions of dense systems of charged "dust" grains in plasmas. Appl. Phys. Lett. **61**(25), 2973–2975 (1992)
42. M. Rubin-Zuzic, G.E. Morfill, A.V. Ivlev, R. Pompl, B.A. Klumov, W. Bunk, H.M. Thomas, H. Rothermel, O. Havnes, A. Fouquet, Kinetic development of crystallization fronts in complex plasmas. Nat. Phys. **2**, 181–185 (2006)
43. A. Melzer, A. Homann, A. Piel, Experimental investigation of the melting transition of the plasma crystal. Phys. Rev. E **53**(3), 2757–2766 (1996)
44. R.A. Quinn, C. Cui, J. Goree, J.B. Pieper, H. Thomas, G.E. Morfill, Structural analysis of a coulomb lattice in a dusty plasma. Phys. Rev. E **53**(3), R2049–R2052 (1996)
45. C.A. Knapek, D. Samsonov, S. Zhdanov, U. Konopka, G.E. Morfill, Recrystallization of a 2D plasma crystal. Phys. Rev. Lett. **98**, 015004 (2007)
46. V. Nosenko, S.K. Zhdanov, A.V. Ivlev, C.A. Knapek, G.E. Morfill, 2D melting of plasma crystals: equilibrium and nonequilibrium regimes. Phys. Rev. Lett. **103**(1), 015001 (2009)
47. J.E. Allen, Probe theory: the orbital motion approach. Phys. Scripta **45**, 497–503 (1992)
48. J.E. Allen, B.M. Annaratone, U. de Angelis, On the orbital motion limited theory for a small body at floating potential in a Maxwellian plasma. J. Plasma Phys. **63**(4), 299–309 (2000)
49. R.V. Kennedy, J.E. Allen, The floating potential of spherical probes and dust grains. II: Orbital motion theory. J. Plasma Phys. **69**(6), 484–506 (2003)
50. C. Cui, J. Goree, Fluctuations of the charge on a dust grain in a plasma. IEEE Trans. Plasma Sci. **22**(2), 151–158 (1994)
51. J.E. Daugherty, R.K. Porteous, M.D. Kilgore, D.B. Graves, Sheath structure around particles in low-pressure discharges. J. Appl. Phys. **72**(9), 3934–3942 (1992)
52. S.A. Khrapak, A.V. Ivlev, G.E. Morfill, Momentum transfer in complex plasmas. Phys. Rev. E **70**, 056405 (2004)
53. J. Goree, Charging of particles in a plasma. Plasma Sour. Sci. Technol. **3**, 400–406 (1994)
54. S.A. Khrapak, S.V. Ratynskaia, A.V. Zobnin, A.D. Usachev, V.V. Yaroshenko, M.H. Thoma, M. Kretschmer, H. Höfner, G.E. Morfill, O.F. Petrov, V.E. Fortov, Particle charge in the bulk of gas discharges. Phys. Rev. E **72**, 016406 (2005)
55. V.E. Fortov, A.P. Nefedov, O.S. Vaulina, A.M. Lipaev, V.I. Molotkov, A.A. Samaryan, V.P. Nikitskiǐ , A. I. Ivanov, S.F. Savin, A.V. Kalmykov, A.Ya. Soloviev, P.V. Vinogradov, Dusty plasma induced by solar radiation under microgravitational conditions: an experiment on board the Mir orbiting space station. JETP **87**(6), 1087–1097 (1998)
56. B. Walch, M. Horanyi, S. Robertson, Charging of dust grains in plasma with energetic electrons. Phys. Rev. Lett. **75**(5), 838–841 (1995)
57. S. Hamaguchi, R.T. Farouki, Polarization force on a charged particulate in a nonuniform plasma. Phys. Rev. E **49**, 4430–4441 (1994)

58. J.P. Boeuf, C. Punset, Charging Time. in *Dusty Plasmas: Physics, Chemistry and Technological Impacts in Plasma Processing*, Chap. 1.1.4, ed. by A. Bouchoule (Wiley, Chichester, 1999)

59. U. Konopka *Wechselwirkungen geladener Staubteilchen in Hochfrequenzplasmen*. PhD thesis, Fakultät für Physik und Astronomie der Ruhr-Universität-Bochum (2000)

60. T. Matsoukas, M. Russell, Particle charging in low-pressure plasmas. J. Appl. Phys. **77**(9), 4285–4292 (1995)

61. U. Konopka, G.E. Morfill, L. Ratke, Measurement of the interaction potential of microspheres in the sheath of a rf discharge. Phys. Rev. Lett. **84**, 891–894 (2000)

62. M. Lampe, G. Joyce, G. Ganguli, Interactions between dust grains in a dusty plasma. Phys. Plasmas **7**(10), 3851–3861 (2000)

63. S.A. Khrapak, G.E. Morfill, Grain surface temperature in noble gas discharges: refined analytical model. Phys. Plasmas **13**, 104506 (2006)

64. A. Melzer, V.A. Schweigert, A. Piel, Transition from attractive to repulsive forces between dust molecules in a plasma sheath. Phys. Rev. Lett. **83**(16), 3194–3197 (1999)

65. J.P. Boeuf, C. Punset, Forces Acting on Dust Particles. in Dusty Plasmas: *Physics, Chemistry and Technological Impacts in Plasma Processing*, chap. 1.2, ed. by A. Bouchoule (Wiley, Chichester, 1999)

66. J.E. Daugherty, R.K. Porteous, D.B. Graves, Electrostatic forces on small particles in low-pressure discharges. J. Appl. Phys. **73**(4), 1617–1620 (1993)

67. S.A. Khrapak, A.V. Ivlev, G.E. Morfill, H.M. Thomas, Ion drag force in complex plasmas. Phys. Rev. E **66**, 046414 (2002)

68. V. Yaroshenko, S. Ratynskaia, S. Khrapak, M.H. Thoma, M. Kretschmer, H. Höfner, G.E. Morfill, Determination of the ion-drag force in a complex plasma. Phys. Plasmas **12**, 093503 (2005)

69. M. Chaudhuri, S.A. Khrapak, G.E. Morfill, Ion drag force on a small grain in highly collisional weakly anisotropic plasma: Effect of plasma production and loss mechanisms. Phys. Plasmas **15**, 053703 (2008)

70. S.A. Khrapak, G.E. Morfill, Basic processes in complex (dusty) plasmas: charging, interactions, and ion drag force. Contrib. Plasma Phys. **49**(3), 148–168 (2009)

71. P.S. Epstein, On the resistance experienced by spheres in their motion through gases. Phys. Rev. **23**(6), 710–733 (1924)

72. B. Liu, J. Goree, V. Nosenko, L. Boufendi, Radiation pressure and gas drag forces on a melamine-formaldehyde microsphere in a dusty plasma. Phys. Plasma **10**(1), 9–20 (2002)

73. H. Rothermel, T. Hagl, G.E. Morfill, M.H. Thoma, H.M. Thomas, Gravity compensation in complex plasmas by application of a temperature gradient. Phys. Rev. Lett. **89**(17), 175001 (2002)

74. S. Mitić, R. Sütterlin, A.V. Ivlev, H. Höfner, M.H. Thoma, S. Zhdanov, G.E. Morfill, Convective dust clouds driven by thermal creep in a complex plasma. Phys. Rev. Lett. **101**, 235001 (2008)

75. M. Schwabe, M. Rubin-Zuzic, S. Zhdanov, A.V. Ivlev, H.M. Thomas, G.E. Morfill, Formation of bubbles, blobs, and surface cusps in complex plasmas. Phys. Rev. Lett. **102**(25), 255005 (2009)

76. M. Rubin-Zuzic, H.M. Thomas, S.K. Zhdanov, G.E. Morfill, Circulation dynamo in complex plasma. New J. Phys. **9**, 39 (2007)

77. S. Hamaguchi, R.T. Farouki, D.H.E. Dubin, Triple point of Yukawa systems. Phys. Rev. E **56**, 4671–4682 (1997)

78. H. Ikezi, Coulomb solid of small particles in plasmas. Phys. Fluids **29**(6), 1764–1766 (1986)

79. K. Kremer, M.O. Robbins, G.S. Grest, Phase diagram of Yukawa systems: model for charge-stabilized colloids. Phys. Rev. Lett. **57**(21), 2694–2697 (1986)

80. E.J. Meijer, D. Frenkel, Melting line of Yukawa system by computer simulation. J. Chem. Phys. **94**(3), 2269–2271 (1990)

81. M.J. Stevens, M.O. Robbins, Melting of Yukawa systems: a test of phenomenological melting criteria. J. Chem. Phys. **98**(3), 2319–2324 (1993)

82. S. Hamaguchi, R.T. Farouki, D.H.E. Dubin, Phase diagram of Yukawa systems near the one-component-plasma limit revisited. J. Chem. Phys. **105**(17), 7641–7647 (1996)
83. O. Vaulina, S. Khrapak, G. Morfill, Universal scaling in complex (dusty) plasmas. Phys. Rev. E **66**, 016404 (2002)
84. O.S. Vaulina, S.V. Vladimirov, O.F. Petrov, V.E. Fortov, Criteria of phase transitions in a complex plasma. Phys. Rev. Lett. **88**(24), 245002 (2002)

Chapter 3
Experiments

The experiments performed in the scope of this thesis aim to contribute to the understanding of the description of a two-dimensional system of particles on the kinetic level, particularly with regard to the phase state.

For this purpose, a two-dimensional system of dust particles had to be generated in the laboratory and its state had to be identified not only by means of structural properties, but in consideration of the particle motion itself. The behavior of those dynamical and structural characteristics during a continuous change of the state of the system has to be observed to understand the underlying processes.

The basic experimental setup is described in the following Sect. 3.1, including the techniques for particle detection and the methods for manipulation of the particle system necessary to obtain a change in its state.

An overview on the data acquisition and primary data analysis to obtain particle coordinates and velocities is given in Sect. 3.2. The uncertainties to be considered in this procedures are addressed in the last Sect. 3.3.

3.1 Experimental Setup

The experiments have been performed in the vacuum chamber shown in Fig. 3.1. The chamber was build into a metal frame which also contained magnet coils (partly shown in the top part of Fig. 3.1) for particle manipulation, and a water cooling system (yellow and blue tubes) for the rf electrode. The chamber itself is an octagon with an alternating edge length of ≈ 193 and 85 mm and a height of 120 mm.

The specific experimental setup used in the work presented here is illustrated in the sketch in Fig. 3.2. The long chamber edges could be used as side windows, while on the short edges and in the top plate there were vacuum flanges for mounting equipment inside the chamber. Two of the side windows were optically accessible, and the central part of the top chamber plate was replaced by a large circular window for viewing from the top with the camera.

C. A. Knapek, *Phase Transitions in Two-Dimensional Complex Plasmas*,
Springer Theses, DOI: 10.1007/978-3-642-19671-3_3,
© Springer-Verlag Berlin Heidelberg 2011

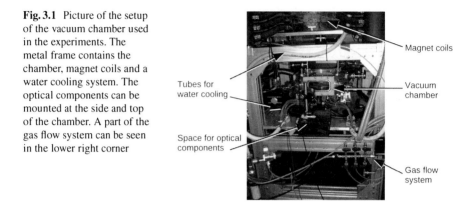

Fig. 3.1 Picture of the setup of the vacuum chamber used in the experiments. The metal frame contains the chamber, magnet coils and a water cooling system. The optical components can be mounted at the side and top of the chamber. A part of the gas flow system can be seen in the lower right corner

Details of the setup shown in Fig. 3.2 will be explained in the following Sects. 3.1.1– 3.1.4.

3.1.1 Vacuum System

The gas inflow was controlled by a thermal mass flow controller with a maximum flow rate of 10 sccm and an accuracy of $\pm 1\%$ of the full scale, calibrated for N_2. Since in all measurements Argon was used as gas, the read out of the flow controller has to be multiplied by the gas correction factor of 1.39 for Ar. The gas was directed into the chamber through a fitting in the top flange, and then pumped out of the chamber by a turbomolecular pump attached to a backing pump. The outlet to the pumps, provided with a valve, was mounted on the side flange directly below the gas inflow to ensure that there were no direct gas flows through the whole chamber during an experiment, which would otherwise influence the particle dynamics.

A Baratron (capacitance manometer) with a range of 13.3 Pa and an accuracy of $\pm 0.25\%$ measured the neutral gas pressure in the chamber and sent the value to a control unit where it was displayed. The combination transducer (Ionivac) shown in Fig. 3.2 is a combination of a hot cathode ionization sensor and a Pirani sensor. It covers a much wider range of pressures ($5 \times 10^{-8} - 10^5$ Pa) but is dependent on the sort of gas and was not used during experiments in favor of the more exact readings at low pressures provided by the Baratron.

3.1.2 Plasma Generation

An electrode was mounted horizontally on the bottom of the chamber, 70 ± 0.6 mm apart from the top plate of the vacuum chamber. It was a flat aluminum disk with

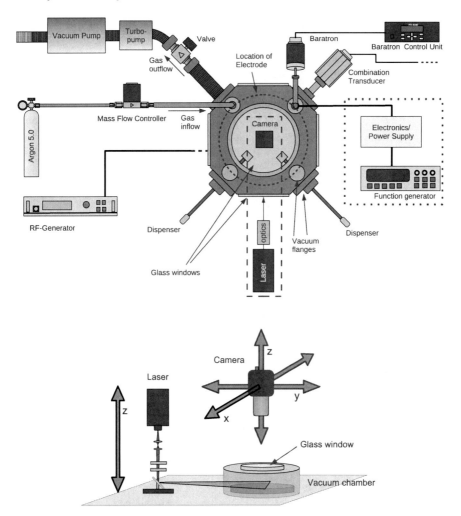

Fig. 3.2 *Top* Sketch of the experimental setup including the chamber (*center*), the vacuum system (*top left*), the gas flow and pressure control units (*left* and *top right*), the RF generator (*bottom left*), the optical components (*dashed box*) and the electronics for particle manipulation (*dotted box*, see Fig. 3.3). *Bottom* Detailed sketch of the optical system used for particle illumination and recording (devices enclosed by the *dashed box* in the upper sketch). The laser beam is directed into the chamber through a side window as a vertically thin, but horizontally expanded layer, while the camera views from the top. The *thick outlined arrows* indicate the axis along which devices can be moved

a diameter of 196 ± 0.2 mm and a 6 ± 0.2 mm wide elevated rim with a height of 2 ± 0.05 mm for plasma potential shaping.

The electrode was capacitively coupled to a radio-frequency (rf) generator and isolated from the rest of the metal chamber which was grounded and served as the counter electrode. This setup is also called an asymmetric rf discharge (see e.g. [1]), and is shortly explained in the following:

When a plasma is ignited by applying a rf power to the electrode, electrons and ions stream to the surrounding surfaces. Due to the higher mobility of the electrons, the surfaces obtain a negative charge compared to the plasma potential. Potential differences V_{driven} across a boundary layer (sheath) between electrode surface and the bulk plasma, and $V_{grounded}$ across the layer between grounded surface and bulk plasma, appear. While the electrons quickly flow off the grounded parts, the capacitor between electrode and rf generator, which is conductive for the rf frequency, prevents the electrode to discharge. A permanent negative dc self bias V_{SB} is generated. The quantity of V_{SB} depends on the time averaged voltage drops across the sheaths, $V_{SB} = -(\bar{V}_{driven} - \bar{V}_{grounded})$. Since the area of the driven electrode is much smaller than the area of the grounded chamber walls, and the voltage drop is inversely proportional to those areas (the sheath can be seen as a capacitor), $\bar{V}_{driven} > \bar{V}_{grounded}$ and therefore $V_{SB} < 0$.

The rf generator was operated at 13.56 MHz with a maximum power output of 300 W. It further contained a matching circuit to adjust forward and reflected power, and displayed both the adjusted power and the self bias voltage at the electrode. The peak-to-peak voltage V_{pp}—the voltage drop between the driven electrode and the grounded chamber—was measured later with an oscilloscope. V_{pp} is of special interest for repeating the experiment in another setup, or comparing it with other experiments, since the impact on the plasma parameters of setting a certain rf power depends on the geometry of the setup (ie. on the size of the surfaces of chamber and electrode), while V_{pp} is a setup-independent quantity.

3.1.3 Particle Injection and Detection

Two dispensers for dust particles were attached at two side flanges opposite to the gas flow system. The dispensers consisted of a bar extending into the chamber with a small box at the end which contained the dust particles. When the bar was manually shaken from the outside, the particles fell through a sieve out of the box. The dispenser could be moved forward, so that particles could be inserted into the plasma in the middle of the chamber, and back out of the area of interest for examination. The dust particles used in all experiments were melamine-formaldehyde spheres with a diameter of $9.19 \pm 0.09\,\mu m$ and a mass density of $1.51\,g/cm^3$.

A two-dimensional crystal was generated by igniting an Argon plasma at a low pressure (1–2 Pa) and shaking some dust particles into the plasma. The particles quickly obtain a negative charge of the magnitude of several 10^4 e due to the fluxes of electrons and ions to their surface, as it was explained in Sect. 2.1. The particles levitate vertically in the plasma sheath above the electrode where gravity is compensated by the electric force pointing in direction of the bulk plasma for the negatively charged particles. In the horizontal direction, the particles are subject to their mutual, usually repulsive, interaction and to confining forces given by the shaped electrode.

This modifies the radial shape of the plasma potential such that a parabolic potential well is created which confines the particles to the center region at the potential minimum. If the number of particles is small enough (< a few 1000), all can be located in one layer and arrange themselves in a two-dimensional crystalline structure. For large numbers of particles, particles at the edges are pushed into another vertical layer by the radial confining potential and the structure becomes three-dimensional. If this happens, particles can either be dropped down onto the electrode by diminishing the rf power (thus decreasing the levitation height), or, by increasing the power, the highest particle layer can be pumped out of the chamber. This has to be done until a two-dimensional system is obtained, with no particles located below or above the crystal plane.

For optical investigation of the system, the particle layer was illuminated by a 532 nm Nd:YAG laser (TEM00) with a beam diameter <2 mm and a maximum power output of 215 mW. The laser was mounted vertically in front of the chamber (see Fig. 3.2) on an electronic translation slide together with four lenses and a mirror. The first two lenses were spherical biconvex and focussed the beam. The location of the focal point was defined by the distance between those lenses. Two cylindrical planoconvex lenses then spread the beam in one direction. A mirror directed the resultant (horizontally wide, but vertically thin) sheet of light into the chamber.

The Gaussian vertical beam width was roughly estimated to be ≈ 200–$300\,\mu$m in the center of the chamber. This was obtained by scanning vertically through a layer of particles and taking the range in which the particles were still visible. A more accurate measurement for the same setup was done in [2] with a 3D-scanning system consisting of tiltable mirrors. It gave a full width at half maximum of $137\,\mu$m.

The whole optical system consisting of laser, lenses and mirror could be moved in the vertical direction through a computer interface in steps of 0.1 mm for vertical scans through the chamber, and to adjust the laser sheet height to the height of the particle layer. According to the Mie theory for the scattering of electromagnetic waves by spherical particles of a size of the magnitude of the wavelength [3], the laser light is scattered by the grains into all directions and can be recorded by a camera mounted perpendicular to the plane of incidence of the laser beam.

A high speed CMOS camera (Photron Fastcam) was mounted on three translation stages on top of the chamber looking through the top glass window providing images of the plane of particles (Fig. 3.2). It could be moved electronically in all directions, and it was possible to synchronize the vertical motion of the camera with that of the laser. This ensured that the focus of the camera always stayed adjusted to the laser illumination plane. An object lens with a focal length of 105 mm and an aperture of 1 : 2.8 D was used in the experiments.

The camera chip had 1024×1024 square pixels with a pixel size of $17.5\,\mu$m. It recorded time series of a maximum length of 6144 frames per run, with a maximum frame rate of 2000 fps at the full spatial resolution. Usually, frame rates ≤ 500 fps are preferable, since the exposure time at faster recordings is too small to provide reliable particle images bright enough for the analysis. The maximum number

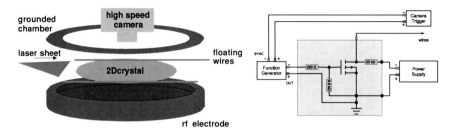

Fig. 3.3 *Left* Sketch of the experimental setup with the wires for particle manipulation. *Right* Circuit for the pulse generation and for synchronizing the pulse with the start of the recording of the camera (corresponding to the devices in the dotted box in Fig. 3.2)

of images and therefore the total recording time, was restricted by the storage space on the memory chip on which the images were stored.

The camera was able to handle trigger signals to match the start of a record with an event like the inducing of the melting of the crystal, as will be described in the next section.

3.1.4 Particle Manipulation

The basic aim of this work is to observe the dynamics of phase transitions in two-dimensional complex plasmas. A possibility to manipulate the state of the particle system is to apply a negative electric pulse to wires mounted inside the chamber. The negatively charged particles are repelled by such a pulse, and if the generated electric force is strong enough to exceed the force responsible for the mutual interparticle repulsion, the initially formed lattice compound breaks open.

For this manipulation, two thin tungsten wires with a diameter of 0.1 mm were mounted parallel and horizontally above the electrode, as sketched in the left panel of Fig. 3.3. The wires could be moved in the vertical direction by a mechanical system mounted through a vacuum fitting, which was manually operated from the outside. Their height was adjusted in each experiment to be at approximately the particle layer height. The gap between the wires was centered over the electrode center and aligned along the direction of the laser sheet. The exact size of the gap was measured for each experiment separately, since it varied within a small range with the wire height.

Both wires were attached to a power supply with a maximum output voltage of 300 V DC via a circuit which was controlled by a function generator. The circuit is shown on the right side in Fig. 3.3. It consists basically of a MOSFET transistor which opens the connection between the power supply and the wires only for the duration of a suitable signal from the function generator. In principle this was a 5 V peak-to-peak square wave signal. The carrier frequency, burst rate and duty cycle set at the function generator defined the occurence and duration of pulses. The voltage at the power supply had to be set to a negative value < -200 V to achieve melting. To avoid repeated melting of the crystal and thus destabilize it, the power supply was turned down except for the time when a recording was done. If not used for the excitation, the wires acquired a negative floating potential in the plasma and contributed to the confinement in the particle plane.

This setup provided the possibility to apply electric pulses simultaneoulsy to both wires for particle manipulation. The SYNC output of the function generator sent a TTL 'low' signal for the duration of a burst and was attached to the camera to provide a trigger signal for the start of a recording.

3.2 Particle Tracking and Tracing

Particle positions were extracted from the images by an intensity weighted center-of-mass method, also called moment method. One image is searched for pixels with intensity values I above a chosen threshold. If one such pixel is found, close-by pixels are checked for their brightness consecutively until the magnitude of the intensity falls below the threshold. If the total number of adjacent pixels with intensity values above the threshold is larger than a chosen minimum, this region is registered as a particle. The particle position (x, y) is calculated as the intensity weighted center of the region:

$$x = \frac{\sum_{i=1}^{n_x} x_i I_i}{\sum_{i=1}^{n_x} I_i}, \quad y = \frac{\sum_{i=1}^{n_y} y_i I_i}{\sum_{i=1}^{n_y} I_i} \qquad (3.1)$$

The sums run over the number of illuminated pixels of one particle in x- and y-direction, n_x and n_y, respectively.

After all images of a time series are tracked for particles, velocities are obtained by tracing individual particles from frame to frame. The tracing algorithm takes a position $\mathbf{x}_k(t) = \begin{pmatrix} x_k(t) \\ y_k(t) \end{pmatrix}$ of a particle k in the frame at time t, and searches in the consecutive frame at time $t + \Delta t$ for a particle within an adjustable radius centered around $\mathbf{x}_k(t)$. Δt is the time step between the consecutive frames. If a particle is found in the specified region, it is assigned to be the same particle k. The search radius has to be chosen large enough according to the expected velocity in order to not loose track of particles, but much smaller than the interparticle distance to avoid assigning a next neighbor to a particle by mistake (this is important if a particle vanishes out of the image, e.g. through vertical displacement; then the algorithm has to stop tracing this particle trajectory).

The velocity is then calculated as

$$\mathbf{v}_k(t) = \frac{\mathbf{x}_k(t + \Delta t) - \mathbf{x}_k(t)}{\Delta t} \qquad (3.2)$$

This procedure is consecutively repeated for all frames of the time series.

3.3 Treatment of Measurement Uncertainties

The accuracy of the tracked particle positions is determined by two factors: the accuracy of the tracking algorithm itself, and the pixel-noise.

The tracking algorithm has to work with a discrete set of intensity values (the pixels) onto which the continuous intensity distribution of the particles is mapped. The deviation of a tracked position from the real center of an intensity distribution then depends on the number of illuminated pixels per particle and on the position of the real center within a pixel (sub-pixel position).

The second source of uncertainties is the pixel-noise. The pixel-noise originates from the finite temperature of the chip of the camera. The chip is a device consisting of semiconductor cells (pixels) in which incident photons create electron-hole pairs. A finite temperature causes vibrations of the atoms in the pixels and randomly generates electron-hole pairs by inelastic collisions, even if the chip is not exposed. This pretends incident photons when the pixels are read out. In the image it appears as random intensity values drawn from a Gaussian distribution with parameters depending on the quantity of thermal vibrations, which are superimposed on each pixel and falsify the intensity profile of particle images. Due to the intensity weighting in the tracking procedure, in the further course this causes a misinterpretation of the tracked particle position.

The full procedure for the investigation of both error sources together with detailed results is given in Chap. 9. A short summary is given in the following paragraph: Sets of ten images, each containing 2500 artificial particles have been generated. An artificial particle was defined by a two-dimensional Gaussian intensity profile, with the mean being the "real" particle center position and the width being the particle size. The real particle centers in the first image were located on the nodes of a square grid representing the pixel grid, plus a small random component which placed the centers on arbitrary sub-pixel positions. Particle displacements from frame to frame were then defined as random values drawn from a Gaussian distribution, which were added to the coordinates of each particle in the image to obtain the next image. This simulates a random particle vibration as one would expect for Brownian motion of particles with a finite temperature. This procedure was then iterated to obtain ten images.

To examine the effects of particle image size (the size the particle appears to have in units of pixels), 17 different Gaussian widths of the intensity profile have been chosen to obtain particle sizes from 1 to 33 pixels per particle. Also, 14 different quantities of particle displacements from frame to frame were tested, by picking 14 values from the range 0.014 to 0.285 pixels as the width of the random displacement distribution.

All sets of images produced so far were copied and additionally superimposed with artificial pixel-noise. For each image, a separate noise matrix of the size of the image matrix, containing random numbers, was generated. The noise matrix was added to the image and the resulting image was scaled back to the original color space. The random numbers where again taken from Gaussian distributions with

10 different widths—simulating different noise levels—from the interval 2 to 20 in units of intensity (the maximum intensity is 255).

In total, that gives $17 \times 14 \times 10 = 2380$ sets of 10 images each for analysis of the noisy data, and $17 \times 14 = 238$ sets without noise (the original images) to investigate solely the quality of the tracking algorithm.

The images have been fed into the same procedure as the experimental data to obtain particle coordinates and velocities, or in this case displacements from frame to frame. Since each manipulation of the originally defined particle position had been stored as the 'real' particle coordinates, the differences between the original values and the estimated values could then be analyzed statistically.

Particle Coordinates The results in the case of zero noise show a systematic error depending on the respective sub-pixel position, in particular for small particles. This effect is often referred to as pixel-locking: The resolution of a narrow intensity profile is very coarse, and small movements are not imaged one-to-one on the pixel grid. The particle seems to move much slower in some parts, or even stand still, while it practically jumps from one position to the next in another part. For particle sizes >10 pixels, the dependence on the sub-pixel position decreases quickly and the error becomes more uniformly distributed within a pixel (for an example, see Figs. 9.2 and 9.3 in the appendix).

The imposed pixel-noise changes the situation qualitatively. The formerly systematic errors are masked by a larger statistical error even for small particles ≥ 3 pixels if the width of the noise distribution is larger than 10. For particles with more than 9 pixels any noise level causes the total error to become statistical. Fig. 3.4a–c shows the most probable error (the highest peak in the histogram of absolute error values), the maximum error and the mean error (standard deviation of the error distribution) vs. noise level for particle sizes of 2–11 pixels. The most probable error is shown, because in most cases the maximum error appears for a very small number of sub-pixel positions only, and therefore gives a misleading impression of the quantity of the error one has to expect. The mean error only makes sense for statistical error distributions.

The results can be interpreted in the following way: For noiseless data (noise level 0 in Fig. 3.4) and for particle image sizes smaller than three pixels regardless of the noise level, the most probable error should be chosen. This error is also applicable for particle image sizes ≥ 3 pixels and noise levels <10. All other data have a superficial statistical error, and the standard deviations presented in Fig. 3.4c can be used.

One can simplify this situation, since from the noise-level 4 on the width of a Gaussian fit to the error distribution represents the quantity of the error reasonably well for particles with more than six pixels (see Chap. 9), though one has to keep in mind that the error in that cases is not perfectly statistical. In general, at noise level 0 the most probable error drops fast from a maximum of 0.3 pixels for particles with 2 pixels, to 0.01 pixels for particles consisting of 11 pixels. This trend is also visible for increasing noise, while at the same instant, with increasing noise the error decreases for equal numbers of pixels per particles. When the error becomes statistical, it increases slightly with increasing noise, whereas the dependence on the particle size is small.

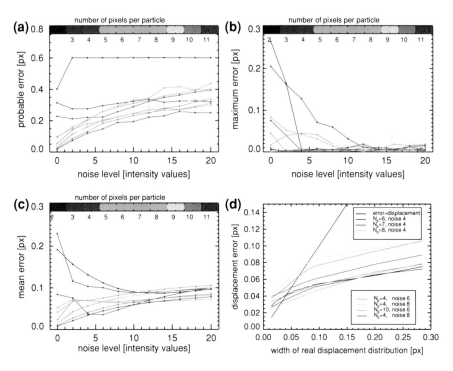

Fig. 3.4 Error of absolute particle positions vs. noise level for particle sizes of 2–11 pixels/particle (color-coded). **a** Most probable error, position of the maximum of the histograms of absolute error values. **b** Maximum error. **c** Standard deviation of a Gaussian fit to the error histograms. **d** Error of particle displacements vs. the width of the real particle displacement distribution for selected cases of particle size and noise level (see legend in the plot). The *black solid line* is the bisector where the size of the error is equal to the width of the displacement distribution. The presented curves are relevant for most experiments presented in the thesis

Note that the absolute particle positions have no physical importance in most of the analysis carried out. Mostly distances between particles, or displacements of a particle from frame to frame are of interest. The systematic error in the coordinates therefore has not much impact.

Particle Displacements The implication of a systematic error δx on the errors of the particle displacements is a dependence of the error on the actual distance a particle moves: the errors have a direction and can not be handled by error propagation, rather they are added or subtracted, respectively, if a distance $\Delta x = x_2 - x_1$ is calculated:

$$\Delta x + \delta(\Delta x) = x_2 + \delta x_2 - (x_1 + \delta x_1) = (x_2 - x_1) + (\delta x_2 - \delta x_1) \qquad (3.3)$$

For close-by positions it was found that the errors often are very similar and have the same sign. Then $\delta x_2 - \delta x_1$ can become very small.

To get an idea of the behavior of the error of displacements from frame to frame, it was calculated directly from the deviation of the original, real displacement and the respective displacement estimated by the tracing algorithm. It was found that the displacement errors have no systematic features but are Gaussian distributed for all cases, with and without noise. The Gaussian width depended on the quantity of the particle displacement (representing for example a particle temperature), the particle size and the noise level. Therefore, the error in the displacements is the standard deviation of the error distribution which is shown in Fig. 3.4d for selected cases. In all examples, the error increases considerably with increasing particle displacement and with decreasing particle size.

The dependence of the error on the actual quantity of particle displacements complicates the error analysis for e.g. velocities from experimental data: The size of the uncertainty is only known, if the magnitude of the velocity is known. The following example, using numbers from experimental data, should illustrate the implications of the displacement error: Assuming particles at room temperature ($T = 293.15$ K) with a mass of $m = 6.14 \times 10^{-13}$ kg and a spatial resolution of 6.74×10^{-3} mm/px, and assuming a Maxwellian velocity distribution, the width of this distribution would be $\sigma_v = \sqrt{k_B T/m} \approx 12$ px/s. With a frame rate of 500 fps this would be a displacement width of 0.024 px. According to Fig. 3.4d, a possible error in the displacements for low pixel-noise is ≈ 0.03 px ($\hat{=} 15$ px/s) in that case. Since velocity distribution and error distribution are Gaussian, the distribution which would actually be measured is given by a convolution of them, yielding an apparent velocity distribution with the width 19.2 px/s $\hat{=}$ 0.064 eV ≈ 746 K.

To find the error in an experimentally obtained data set, the average number of pixels per particle and the noise level must be known. Then a comparison with the artificial particles used above yields an estimate of the uncertainties to be taken into account. The number of pixels per particle is easily obtained during the tracking procedure. The noise level can be extracted from the images by analyzing the histogram of intensity values of the image background between the particles. Depending on the camera settings it can happen that the pixel-noise background is cut off in the dark regions of the images, but a procedure to estimate the noise level from intensity fluctuations of the particle images has been developed for this case (R. Sütterlin, private communication) and is explained in Chap. 9.

The uncertainties in the displacements restrict the resolvable distances in a measurement: Any distance, e.g. the particle motion from frame to frame, has to be much larger than the estimated displacement error to retain a physical meaning. This applies also to distances between particles in one frame, though in this case the error, determined by the difference of the sub-pixel components of the coordinates, becomes small compared with the interparticle distance of usually several pixels. If velocity distributions are to be measured, one has to keep in mind that such a distribution is easily masked by a Gaussian error distribution. Since the distributions are obtained as histograms of velocities, the bin widths should then be chosen to be larger than the expected errors to even out statistical fluctuations caused by the errors in case of small counts per bin.

References

1. Y.P. Raizer, Gas Discharge Physics, chapter 13 (Springer, Berlin, 1991)
2. A. Elsässer, Experimental investigation of complex plasmas by laser tomography. Diploma thesis, Technische Universität München (2006)
3. G. Mie, Beiträge zur Optik trüber Medien, speziell kolloidaler Metallösungen. Ann. Phys. **330**(3), 377–445 (1908)

Chapter 4
Data Analysis Techniques

In the following sections the concepts and techniques used in the data analysis are illustrated. Section 4.1 shortly explains the calculation of charge and screening parameter by wave spectra analysis. The structural properties of the two-dimensional system, and the methods to obtain them are described in Sect. 4.2 starting with the defect analysis in Sect. 4.2.1. The long range translational and orientational order of the system are described by means of the pair- and bond-correlation functions in Sect. 4.2.2 and 4.2.3. A measure for local order is introduced in Sect. 4.2.4 with the bond order parameter which can be defined at each respective particle position within the lattice. The last Sect. 4.3 concludes this section with the statistical description of the dynamics of a system of particles with regard to distribution functions of displacements and velocities.

4.1 Estimation of Particle Charge and Screening Parameter

The particle charge Q and the screening parameter $\kappa = \Delta / \lambda_D$ can be derived from the sound velocities of phonons in a plasma crystal. Here Δ is the interparticle spacing and λ_D the Debye length of the particles. Phonons are natural waves which exist in any plasma crystal due to random particle motions without the need of excitation. Their wave numbers k and the wave frequencies ω for longitudinal (index "l") and transverse (index "tr") waves are connected through the dispersion relations $\omega = c_{l,tr} k$, with the respective sound velocities c_l and c_{tr}. The following relations show the dependence of the sound velocities on the charge and κ in the limit of $k \rightarrow 0$ and $\kappa \rightarrow 0$ [1, 2]:

$$\frac{c_{tr}}{c_0} = 0.51317 - 0.0226(\kappa / l_{\mathrm{corr}})^2 \approx 0.51317 \qquad (4.1)$$

$$\frac{c_l}{c_0} = \frac{2.585}{\sqrt{\kappa / l_{\mathrm{corr}}}} \qquad (4.2)$$

C. A. Knapek, *Phase Transitions in Two-Dimensional Complex Plasmas*,
Springer Theses, DOI: 10.1007/978-3-642-19671-3_4,
© Springer-Verlag Berlin Heidelberg 2011

$$c_0 = \Omega_0 \Delta = \frac{Q}{\sqrt{4\pi \epsilon_0 m \Delta^3}} \Delta \qquad (4.3)$$

Ω_0 is the plasma frequency and $l_{corr} = \sqrt{2/\sqrt{3}} \approx 1.075$ is a correction factor accounting for the geometrical arrangement in a 2D hexagonal lattice.

With c_0 the particle charge can be calculated as

$$Q = c_0 \sqrt{4\pi \epsilon_0 m \Delta} \qquad (4.4)$$

The screening parameter κ is given by the direct dependency of c_l on $(\sqrt{\kappa})^{-1}$ in (4.2). The sound velocities can be obtained from the wave spectra $V(k, \omega)$ of the phonons by the following procedure [3–6]:

The particle velocities $v_{x,y}(t)$ are averaged in spatial bins of the width δx and in temporal bins of the width δt for the directions x and y in the image separately. The bin widths have to be chosen accordingly to get a good resolution. This gives matrices $V(x, t)$, $V(y, t)$ with equally spaced entries—each component (x, t), (y, t) corresponds to an equally spaced range in space and time, with the quantity of the entry being the average of the velocity components of all particles falling into that range.

$V(x, t)$ and $V(y, t)$ are then Fourier transformed to the frequency space to obtain the spectra $V(k_x, \omega)$, $V(k_y, \omega)$. The two matrices for x and y are then squared and added to get the final spectrum $V(k, \omega)$, dependent on the frequency ω and the wave number k. The left panel of Fig. 4.1 shows an example of such a matrix in the frequency space. Each pixel in the plot corresponds to an entry of $V(k, \omega)$ at the respective position (k, ω). The brightness is a measure for the quantity of the matrix entry, where brighter pixels mean higher values. The two wave branches ω_l, ω_{tr} can be seen as the brighter accumulations of pixels.

Both branches of the spectrum have to be fitted by lines for small k as it is implied by (4.1) and (4.2). From the theory follows [1], that the transverse branch ω_{tr} (lower branch in Fig. 4.1), giving the particle charge, is linear over a wider range, while the longitudinal branch is linear only for very small k. Due to the often very poor resolution and contrast at low k, which arises from the limited system size and length of the time series in real space, the fitting of a line to $\omega_l(k)$ is usually subject to large errors. Note that the large uncertainty in the particle velocities discussed in Sect. 3.3 further diminishes the quality of the spectra.

Instead of directly applying fits to the spectra, first the positions $(k, \omega_{l,tr})$ defining the wave branches are identified. For one value of k and a restricted range of ω at approximately the position of one of the wave branches, all pixels brighter than a threshold are chosen. This is done for both branches separately. $\omega_{l,tr}$ is then the intensity weighted center of these pixels. The dispersion of intensity values gives error bars for $\omega_{l,tr}$. The positions $(k, \omega_{l,tr})$ define the sound velocities $v_{l,tr} = \omega_{l,tr}/k$. The obtained sound velocities are plotted vs. k in Fig. 4.1b, with error bars are transformed from the error bars of $\omega_{l,tr}$.

It is now possible to fit the theoretical model $c_{tr} = 0.51317c_0$ to the transverse branch v_{tr} (lower curve in Fig. 4.1b) with c_0 as fit parameter, and with that to obtain

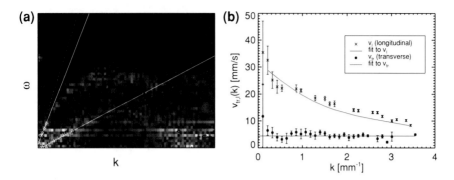

Fig. 4.1 **a** Example of a wave spectrum for one of the performed experiments. Brighter pixels mean a higher density of velocities at the respective wave number k and frequency ω. *Yellow diamonds* mark the points used for the direct linear fit (*white lines*) to the two branches. The *blue* and *red dashed lines* correspond to the fits of the theoretical model. **b** Measured sound velocity vs. k. The error bars are calculated from the dispersion of intensity values at one k from the spectrum in (**a**). The *red* and *blue lines* are fits of theoretical curves giving $\kappa = 0.56 \pm 0.23$ from the longitudinal, and $Q = 10500 \pm 300$ e from the transverse sound velocities, respectively

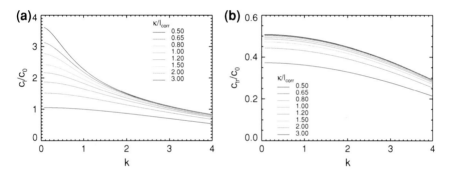

Fig. 4.2 Theoretical curves c_l/c_0 (**a**) and c_{tr}/c_0 (**b**) vs. wave number k for values of $\kappa = 0.5$, 0.65, 0.8, 1, 1.2, 1.5, 2, 3. The wave polarization direction is $0°$

the particle charge Q from (4.4) (the interparticle spacing Δ is usually known from the pair correlation function, see Sect. 4.2.2). The longitudinal branch in that range of k is clearly not linear. To find a function which could be fitted over the whole range of k, theoretical curves $c_{l,tr}/c_0$ for different κ/l_{corr} between 0.5 and 3 were calculated for a similar range of k with the c_0 obtained from the transverse wave. The transverse velocities shown in the Fig. 4.2b are only weakly dependent on κ, as expected. From the longitudinal velocities c_l/c_0 (shown in Fig. 4.2a) a polynomial fit connecting the curves for different κ was used to create coefficients depending on κ. Then a fifth grade polynomial with those coefficients was fitted to v_l/c_0 (red line in Fig. 4.1b), yielding a best estimate for κ.

4.2 Structural Analysis

4.2.1 Defects

Defects are disruptions of the crystal structure. Possible defects in a two-dimensional system are point defects—or disclinations—consisting of a vacancy or an interstitial. The lattice around such an isolated defect is distorted so that the crystal structure is maintained. In a hexagonal lattice with typically six nearest neighbors to each lattice site, the most common point defects are 5-folded (vacancy) or 7-folded (interstitial) lattice sites, one of which is illustrated in Fig. 4.3a.

Point defect positions are found by performing a Delauney triangulation on the particle coordinates (x, y) in an image. In the later analysis, the Triangle-algorithm described in [7] was used. The triangulation covers the two-dimensional xy-surface with a mesh of triangles between neighboring particles under the condition that lines never cross. Each lattice site is then connected by n lines to the adjacent n lattice sites. Those bonds define the number and position of all nearest neighbor particles. In the hexagonal 2D lattice, the lattice site is a point defect if $n \neq 6$.

Dislocations Another type of defect is the dislocation which can be understood as an additional row of particles inserted into an ideal lattice [8]. At the end of this row there will be a pair of disclinations to adjust the lattice, usually a pair of a 5- and 7-fold defect as seen in Fig. 4.3b. In the following the expression "dislocation" will mean a pair of a 5-fold and 7-fold disclination, since this is the most prominent in the hexagonal lattice. Dislocations can form pairs as shown in Fig. 4.3c. This configuration, also called "dislocation pair", is used in the theories for dislocation-mediated two-dimensional melting, which will be introduced later.

Dislocations can be described by means of their Burgers vector [8, 9]. If one draws a closed path around a dislocation, jumping from one lattice site to the next, the same path (the equal number of jumps in the same directions as before) will not close in an ideal lattice. The Burgers vector is the additional vector needed to close that path. It is perpendicular to the dislocation line, i.e. the vector connecting the two disclinations. In the case of a dislocations pair as in Fig. 4.3c, the net Burgers vector will be zero. For this, the two dislocations do not necessarily have to be adjacent in the lattice, only their orientation (the direction of the Burgers vector) is important.

Defect Analysis The arrangement of defects in a lattice, for example as dislocations, can provide valuable information on the system in addition to the absolute number of disclinations. To investigate dislocations, one has to assign adjacent point defects to each other. In experimental data the following defect configurations (for 5- and 7-folds) can usually be seen:

1. isolated, "free" disclinations (either 5- or 7-fold)
2. isolated, "free" dislocations
3. open chains with alternating 5-folds and 7-folds
4. closed chains with alternating 5-folds and 7-folds ("loops", loops of four adjacent defects are an unique dislocation pair)

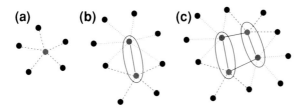

Fig. 4.3 Defects in a two-dimensional hexagonal lattice. **a** Free disclination (5-fold). **b** Free dislocation (pair of a 5- and a 7-fold disclination). **c** Pair of dislocations. The lines indicate the nearest neighbor bonds with *dashed lines* symbolizing 5-fold and *dotted lines* 7-fold lattice sites

5. clusters of defects (a larger amount of disclinations in one region, without apparent structure)

The points 2–5 always contain at least two defect lattice sites directly adjacent to each other. Point 1 does not need further examination, and for the second case the assignation of the defect pair is straightforward. There are two approaches for analysis of the other cases:

1. Always pair those 5- and 7-fold defects with the smallest distance between them. This is a very simple procedure, but it can only identify free dislocations, not complex structures such as chains or loops.
2. After finding a defect pair, search all neighboring lattice sites of each defect for other defects successively until no further adjacent defects are found. With this procedure, chains and closed loops can be identified. Dislocations can be associated within this structures. Defect clusters can not be specified correctly, since they are misinterpreted as chains or loops.

The second procedure is preferable, because it finds both free dislocations and chains or loops. The only problem are the identification of clusters with no apparent structure, or the misinterpretation of randomly distributed disclinations, which often appear in liquid-like states of high disorder, as structures.

Point defects with less than five, or more than seven neighbors can be observed in complex plasmas, but their numbers are very small, especially in the crystalline state they are practically absent.

4.2.2 Pair Correlation Function

The pair—or translational correlation function $g(r)$, also called radial density distribution, shows the probability to find a particle in a distance r from another particle [10]. It is computed for an image by choosing consecutively each particle as a center particle i and counting the number of particles j found in a ring with radius r and width dr around that particle. r goes up to a maximum radius r_{max}. The results for

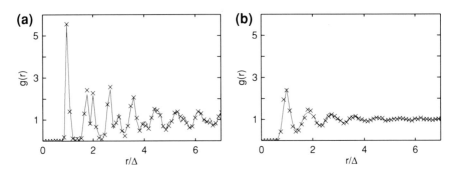

Fig. 4.4 Examples of $g(r)$ found in two-dimensional complex plasmas. r is normalized by the mean particle distance Δ. The *solid lines* correspond to fits using (4.6) and (4.7). **a** Solid state with the fit function proposed by Beresinskii. **b** Liquid state with exponential fit

the different center particles N_{cp} are averaged and then normalized by the particle density $N_{cp}/(\pi r^2)$ times the ring area $2\pi r dr$ for each r. The normalization factor was further improved by taking into account that rings might cross the edge of the analyzed region, or the image edges. Therefore a correction factor ρ_{corr} was calculated specifically for each center particle and r from a simple geometric construction, and the bin counts were normalized by it. This ensured that for large r, $g(r)$ goes to 1.

$$g(r) = \frac{1}{2\pi r dr} \frac{\pi r_{max}^2}{N_{cp}} \frac{1}{N_{cp}} \sum_{i=1}^{N_{cp}} \frac{1}{\rho_{corr,\,i}(r)} \sum_{r-dr<r_j-r_i\,\leq r+dr} 1 \qquad (4.5)$$

The shape of $g(r)$ in an ideal lattice is composed of singular peaks at the distinct distances given by the lattice constant, and of an decaying envelope of the peak amplitudes due to the normalization. For $r \rightarrow \infty$, $g(r)$ goes to 1. In the real crystal, the thermal motion of the particles caused by the finite particle temperature T broadens and lowers the peaks. The measured curve then looks like a series of Gaussian functions, each centered around the respective distance given by the lattice constant (or interparticle distance). The peak amplitudes decrease with increasing r, as shown in the example in Fig. 4.4a. In a liquid-like state, Fig. 4.4b, the peaks become wider and begin to overlap while their amplitude decays very fast exponential with r.

The following fit function for $g(r)$ was proposed in [11] and used in 2D complex plasma analysis [12]; the free fit parameters (k, σ_0, Δ) are highlighted:

$$g_{fit,\,1}(r) = \left[\frac{k}{\sqrt{2\pi}} \frac{1}{\sigma_0} \sum_i \frac{g_{id}(x_i)}{\Delta\, x_i} \exp\left(-\frac{(r-\Delta\, x_i)^2}{2\,\sigma_0{}^2} \right) - 1 \right] \times \exp\left(-r/\xi\right) + 1$$

$$(4.6)$$

with a prefactor k depending on the normalization of $g(r)$ by the particle density, the peak width σ_0, the mean interparticle distance Δ and an exponential decaying envelope $\propto e^{-r/\xi}$. ξ is also called translational correlation length. $g_{id}(x_i)$ is the total number of particles found on a ring with radius x_i around a center particle in the ideal hexagonal lattice. The positions x_i are distinct and defined by the translation vectors $(1, 0)$ and $(1/2, \sqrt{(3)}/2)$ of the lattice. From this, $g_{id}(x_i)$ can be calculated. The summation in (4.6) goes over all calculated x_i, here $n = 120$ positions were used. This function is valid in liquid states, where an exponential decaying $g(r)$ is expected [13]. Another function was introduced by Beresinskii in [14, 15] for solid systems in one and two dimensions. The main difference to (4.6) is a peak width $\tilde{\sigma}$ depending on the distance r. With the correct normalization factors the following fit function was constructed (with the fit parameters $A_{hex}, \sigma_0, \Delta, \xi$):

$$g_{fit,2}(r) = \left[\frac{A_{hex}}{(2\pi)^{3/2}} \frac{1}{\tilde{\sigma}} \sum_{i=1}^{n} \frac{g_{id}(x_i)}{\Delta x_i} \exp\left(\frac{-(r - \Delta x_i)^2}{2\tilde{\sigma}^2} \right) - 1 \right] \times \exp\left(-r/\xi \right) + 1$$

(4.7)

with $\tilde{\sigma} = \sigma_0 \sqrt{\ln \frac{r}{r_0}}$, $r_0 = 0.3 \Delta$

The prefactor is composed of the inverse particle density A_{hex}, which should be equal to the area of one Voronoi cell around a particle in a hexagonal lattice, the factor $1/\sqrt{2\pi}\tilde{\sigma}$ from the Gaussian shape of the peaks, and $1/(2\pi)$ from the normalization of $g(r)$. Additional, a parameter r_0 is introduced, which was theoretically estimated to be between 0.2 and 0.4 for the investigated system. It was chosen to be 0.3 in the fit. The peaks become wider with increasing r in this model, which agrees well with the observations in the experimental data. In fact, this function fitted the data better than (4.6) even in liquid-like states of high disorder. The following paragraph gives some details on the fit procedure itself to explain the interpretation of the outcome of the fit.

Fit Procedure The fit procedure used here is implemented in IDL (Interactive Data Language) and described in [16]. It performs a Levenberg-Marquardt least-squares minimization on a given set of points $y(x)$ using a user-supplied function $f(x, a)$. Two functions were written using (4.6) and (4.7), with the free parameters $a = (A_{hex}, \Delta, \sigma_0, \xi)$. One set of guessed, initial parameters have to be supplied to the fit procedure. The best set of parameters is then the set which minimizes $\sum_x [y(x) - f(x, a)]^2$. In each iteration, the parameters are varied in direction of their negative gradient until the minimum is reached. In general this is a very effective method, but in case of the rather complex function $g_{fit}(r)$ with four free parameters, the choice of the initial starting parameters can become crucial. There is always the danger that local minima of the parameter set appear for certain starting values, which then lead to ambiguous results. Therefore the fit should be repeated with different configurations to ensure the validity of the result.

The fit procedure also provides the 1-σ uncertainty for each fit parameter, which is calculated during the fit as described in [17], and the χ^2 as a measure of the goodness of the fit. χ^2 is the weighted sum of squared distances between data and fit, divided by the squared uncertainties of the data. Often the reduced χ^2 is stated as $\chi_\nu^2 = \chi^2/N_{dof}$ with N_{dof} being the number of degrees of freedom (number of fitted points minus number of fit parameters). The fit procedure assumes that the supplied uncertainties reflect the deviations of the fit model to the real data, therefore only by supplying valid uncertainties a meaningful χ^2 and 1-σ can be obtained. Since the uncertainty in the single points of $g(r)$ is of purely statistical nature (only numbers of particles were counted), and the number of particles used in the analysis is very high (>1000), the statistical error bars are very small, which makes χ^2 very large. With this kind of uncertainties it is not possible to take into account a deviation of the experiment from the model. Therefore, χ^2 should not be interpreted as the real probability of the goodness of the fit, but the ratio $\chi^2_{v,B}/\chi^2_{v,E}$ (index B: Beresinskii fit, index E: exponential fit) can be used as an estimate to choose the best fitting function. To obtain more realistic values for the 1-σ uncertainty of each fit parameter, the reduced χ^2_v can be manually set to 1, implying that the fit is the best possible. With that, the 1-σ uncertainties are calculated again, and should now represent the actual deviation of data to fit function [12].

Interpretation The pair correlation function provides a good estimate for the mean interparticle spacing Δ as the position of the first peak. Further is gives information on the range of translational order in a system, expressed by the correlation length ξ, which is of importance for some established theories of phase transitions, as will be discussed in later chapters.

Aside from the structural information immanent in $g(r)$, the fit parameter σ_0 is correlated to the particle temperature and with that to the dynamics of the particle motion. σ_0 is the dispersion of the particle separation, or lattice constant. This quantity is related to the radius σ_r of the area a single particle occupies in average while it oscillates with the frequency Ω_E around its lattice site (σ_r is the width of the particle displacement distribution):

$$\sigma_r = \sqrt{\frac{k_B T}{m\Omega_E^2}} = \frac{\sigma_0}{\sqrt{2}} \tag{4.8}$$

The factor $\sqrt{2}$ comes from the fact that σ_0 is the width of the Gaussian distribution $f(\Delta_{ik})$ of the vectors Δ_{ik} between two particles i, k. The mean $\bar{\Delta}$ of the values Δ_{ik} is the distance between the mean lattice sites of the two particles, and $\mathbf{r}_{i,k}$ are the respective displacement vectors of the particles i, k from their mean lattice site. Since $\Delta_{ik} - \bar{\Delta} = \mathbf{r}_k - \mathbf{r}_i$, it holds:

$$f(\Delta_{ik}) \propto \exp\left\{-\frac{(\Delta_{ik} - \bar{\Delta})^2}{2\sigma_0^2}\right\} = \exp\left\{-\frac{2\mathbf{r}^2}{2\sigma_0^2}\right\} = \exp\left\{-\frac{\mathbf{r}^2}{2\sigma_r^2}\right\} \propto f(\mathbf{r}_i)f(\mathbf{r}_k)$$
$$\Rightarrow \sigma_0^2/2 = \sigma_r^2$$

In the equality it was used that all $\mathbf{r}_{i,k}$ have the same distribution, which is only shifted in space by a constant factor, therefore $\mathbf{r}_i^2 + \mathbf{r}_k^2 = 2\mathbf{r}^2$, and that the motion of the two particles are uncorrelated. Then $\mathbf{r}_i\mathbf{r}_k \approx 0$. In other words, $f(\Delta_{ik})$ is the convolution of two independent Gaussian distributions. Due to the convolution invariance of Gaussian distributions, the relation between the widths also follows directly.

4.2.3 Bond Correlation Function

An ideal hexagonal crystal has angles of multiples of $60°$ between any two nearest-neighbor bonds, no matter how far the bonds are separated in space. This defines an orientational long range order which can be described by the bond correlation function $g_6(r)$ [13, 18]. To calculate $g_6(r)$, the nearest neighbor bonds have to be identified by a Delauney triangulation. The bond correlation function $g_6(r)$ is then defined as

$$g_6(r) = \left| \frac{1}{N_B} \sum_{l=1}^{N_B} \frac{1}{n(l)} \sum_{k=1}^{n(l)} \exp\{i6(\theta(\mathbf{r}_k) - \theta(\mathbf{r}_l))\} \right| \tag{4.9}$$

Here N_B is the total number of bonds in the crystal, $n(l)$ is the number of bonds at the distance r from bond l, $\theta(\mathbf{r}_{k,l})$ are the respective angles of bonds at $\mathbf{r}_{k,l}$ to an arbitrary axis. Note that $g_6(r)$ is always 1 for the perfect hexagon by definition. In the solid state, $g_6(r)$ should be constant and close to 1. Further, power-law and exponential decays for large r are predicted in hexatic and liquid states, respectively [12, 13]. The hexatic state is an intermediate two-dimensional state assumed to appear between the solid and liquid phase according to some theories. This state will be addressed later in Chap. 6.

The following three models were fitted to $g_6(r)$:

1. Exponential decay $g_6(r) = A_1 e^{-r/\xi_6}$ with the fit parameters A_1 and ξ_6
2. Power-law decay $g_6(r) = A_2 r^{-\eta_6}$ with the fit parameters A_2 and η_6
3. Linear decay $g_6(r) = c_6 r + A_3$ with the fit parameters A_3 and c_6.

The third model, the linear decay, is not mentioned in the theories, but it was added due to the findings in the experimental data.

The fits are applicable for large r only, since the bond correlation function describes the long range behavior of the system. Values of $r < 2–3\Delta$ have therefore to be omitted in the fit. The degree of order goes roughly from unordered (exponential) to ordered (linear). ξ_6 serves as a correlation length comparable to ξ of the pair correlation function. The decay of the power-law is slower than the exponential decay and could be applicable for a better ordered system.

The linear decay can be seen as a practically constant $g_6(r)$, but under the influence of long range effects not considered in the common theories of two-dimensional melting. In all experimental data a substantial slope $c \neq 0$ was found in states appearing rather crystalline with regard to other properties, while power-law decays where hard to find at all. To examine the effect leading to a linear decay, an artificial particle lattice was generated and modified to simulate different deviations from the ideal lattice (compression, domain forming, rotation of domains). The bond correlation function was calculated and a linear decay of $g_6(r)$ appeared in such cases when larger adjacent domains where formed which were rotated relative to one another by large angles, while the unit cells around particles within each domain were kept ideal hexagonal. The domain boundaries consisted necessarily of defect strings to

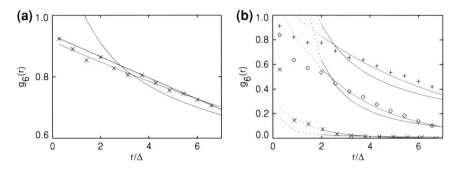

Fig. 4.5 Examples of $g_6(r)$ as found in experimental data of two-dimensional complex plasmas. r is normalized by the mean particle separation Δ. **a** The linear decay (*black line*) fits best, the exponential (*red line*) fits well but has to be omitted because $\xi_6 \approx 26\Delta$ which is larger than the actual system size. The power-law (*blue line*) fits only poorly. **b** Examples of different states of order, distinguished by different plot symbols ($\times, \diamond, +$). Colored lines correspond to fits with the decay power-law (*blue*) and exponential (*red*). The fits were performed for $r > 2\Delta$ only as indicated by the change from dotted lines to solid lines

compensate for the deformation. The complete procedure and graphical results are given in detail in Chap. 10.

This result coincides with the observation in the experimental data, where the same kind of rotated domain structures could be identified. It should be noted that the strong influence of the domains on $g_6(r)$ could mask any power-law or exponential decay as well as prevent $g_6(r)$ from being constant for large r.

Examples for $g_6(r)$ are given in Fig. 4.5 for the two states crystalline (Fig. 4.5a) and liquid-like (Fig. 4.5b). The different models of linear (black line), power-law (blue line) and exponential (red line) decay were fitted and are shown for comparison.

4.2.4 Bond Order Parameter

A useful quantity to examine the lattice in terms of the local orientational order is Ψ_6 which is defined as (following the definition given in [19]):

$$\Psi_{6,k} = \frac{1}{n} \sum_{j=1}^{n} e^{6i\theta_{kj}} = \left| \Psi_{6,k} \right| e^{\{i\phi\}} \tag{4.10}$$

$$\phi = \arctan \left\{ \Im(\Psi_{6,k}) / \Re(\Psi_{6,k}) \right\} \tag{4.11}$$

over the n nearest neighbors of each particle k with θ_{kj} being the angle between the nearest-neighbor-bond of the particles k and j and the x-axis (Fig. 4.6a) The axis can in fact be chosen arbitrary as long as it is fixed, but typically the image x-axis is taken.

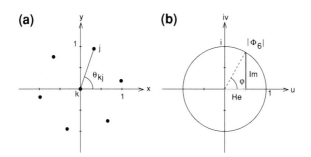

Fig. 4.6 a Hexagonal cell around a particle k defined by its nearest neighbors. The bond between particles k and j has an angle θ_{kj} to the x-axis. **b** Imaginary plane with the *dashed line* being the modulus of $\Psi_{6,k}$ of one center particle k, and the angle ϕ being the argument of $\Psi_{6,k}$. In the ideal hexagon, the modulus is always 1. ϕ lies in the interval $[0, \pi]$ for rotations of the unit cell from 0–$30°$ from the x-axis, and in $[-\pi, 0]$ for angles between 30–$60°$

The modulus $\left|\Psi_{6,k}\right|$ of this complex quantity is the bond order parameter [11] which is 1 by definition for an ideal hexagonal structure. It is often averaged over all particles in the lattice and then used as a measure for the mean local order of the crystal. In Fig. 4.6b it is represented as the length of the dashed line in the complex number plane.

The argument $\phi = \arg(\Psi_{6,k})$ is the angle indicated in Fig. 4.6b. It is a measure for the unit cell orientation with respect to the x-axis (the unit cell is the Wigner-Seitz or Voronoi cell, not the larger cell spanned by the neighboring particles). For rotations of a unit cell around a center particle with respect to the fixed axis, i.e. a common rotation of all nearest neighbors, the angle between the axis and bonds modulo $\pi/3$ is mapped like:

$$(\theta_{kj} \quad \mathrm{mod} \ \pi/3) \in [0, \pi/6] \to [0, \pi]$$
$$(\theta_{kj} \quad \mathrm{mod} \ \pi/3) \in [\pi/6, \pi/3] \to [-\pi, 0]$$

The dependence of ϕ on the degree of rotation is shown in Fig. 4.7. The black dots correspond to the ideal hexagonal unit cell. In the non-ideal crystal the angles might deviate from $60°$, and this dependence is slightly shifted, but still gives an idea of the orientation of the cell. Only for defect lattice sites and strongly deformed hexagons the information in ϕ is not reliable when comparing it to other unit cell orientations.

4.3 Statistical Evaluation of Particle Dynamics

If the crystalline particle system is in local equilibrium, the particle interaction with the plasma is balanced by the neutral gas friction. The particle motion around its mean lattice site can then be described by a Langevin equation [20]

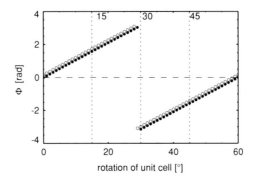

Fig. 4.7 The argument ϕ of Ψ_6 vs. rotation of a unit cell with respect to the x-axis. The *black dots* refer to an ideal hexagon, the *open circles* are a cell with angles between the bonds varying slightly from 60°

$$m\ddot{\mathbf{r}} = -m\Omega_E^2\mathbf{r} - mv_{Ep}\mathbf{v} + \zeta(t) \tag{4.12}$$

with the particle mass m, displacement $\mathbf{r}$ from the mean lattice site and the velocities $\mathbf{v}$. The Einstein frequency Ω_E is the frequency of the particle oscillation around its equilibrium position [9], and v_{ep} the Epstein drag coefficient. $\zeta(t)$ is a stochastical force which is the driving thermal force originating from the finite temperature T of the particles. It causes the particles to perform a random Brownian motion and counteracts the damping and the restoring forces.

The first term on the right hand side of (4.12) describes the restoring force, which drives the particle towards its mean lattice site, as a repulsive electric force between the equally charged particles. A mean lattice site with respect to the surrounding particles is defined as the position of the minimum of the electric potential of all neighboring particles. A particle oscillates around this center with a frequency Ω_E, depending on the shape of the potential and the particle charge. The basic interaction potential for the complex plasma was introduced as the Yukawa-type potential $\Phi(r) = Q_D e^{-r/\lambda_D}/(4\pi\epsilon_0 r)$ with the screening length λ_D, in Sect. 2.1.

The particle motion is damped mainly by collisions with neutral gas atoms. This process is described by the second rhs term in (4.12) and was explained in Sect. 2.2 as well approximated by Epstein damping. The rate of collisions and therefore the rate of damping, v_{Ep}, depends on the neutral gas properties pressure p, gas temperature T_g, mass of the gas atoms m_g and on dust particle properties and can be calculated as [3]:

$$v_{Ep} = \delta\sqrt{\frac{8m_g}{\pi k_B T_g}}\frac{p}{\rho r_p} \tag{4.13}$$

with the radius and mass density r_p and ρ of the particles and the Boltzmann constant k_B. The coefficient δ depends on the mechanism of the reflection of gas atoms from the particle surface. For thermal nonconductive, spherical particles, Epstein calculated $\delta = 1.442$ in the case of diffuse reflection [21]. The coefficient δ was also measured in experiments from horizontal oscillations of particles in a potential well to be 1.48 ± 0.05 [22], and 1.26 ± 0.13 [23]. A vertical resonance method yielded 1.44 ± 0.19 [23].

The Langevin equation of motion can be solved by the Fokker-Planck equation which yields a particle ensemble that obeys a Maxwell-Boltzmann distribution at a particle temperature T. The Hamiltonian for one cell in the lattice is

$$H = E_{kin}(\mathbf{v}) + E_{pot}(\mathbf{r}) = \frac{1}{2}m(\mathbf{v} - \langle \mathbf{v} \rangle)^2 + W(\mathbf{r}) \tag{4.14}$$

The mean value $\langle \mathbf{v} \rangle$ is subtracted to eliminate contributions to the kinetic energy due to motions of the center of the lattice site itself. The probability distribution becomes

$$P(\mathbf{r}, \mathbf{v}) = C \exp\left\{-\frac{H}{k_B T}\right\} = C \exp\left\{-\frac{m(\mathbf{v} - \langle \mathbf{v} \rangle)^2}{2k_B T}\right\} \exp\left\{-\frac{m\Omega_E^2 \mathbf{r}^2}{2k_B T}\right\} \tag{4.15}$$

with C a constant factor depending on the normalization.

The probability distribution can be separated into functions for the several components of the displacement and velocity vectors as long as they are independent:

$$P(\mathbf{r}, \mathbf{v}) = p(\mathbf{r})p(\mathbf{v}) = p(r_x)p(r_y)p(v_x)p(v_y) \propto e^{-\frac{r_x^2}{2\sigma_r}} e^{-\frac{r_y^2}{2\sigma_r}} e^{-\frac{(v_x - \bar{v}_x)^2}{2\sigma_v}} e^{-\frac{(v_y - \bar{v}_y)^2}{2\sigma_v}} \tag{4.16}$$

The above distribution functions are provided directly by the measurement of particle coordinates in the images, provided there is a high enough spatial and temporal resolution to resolve the particle oscillation amplitude as well as the velocity. Of importance here is the quantity of the measurement error which puts a limit on the smallest measurable distance.

The calculation of the velocity was given in (3.2). The displacements $\mathbf{r}(t) = \begin{pmatrix} r_x(t) \\ r_y(t) \end{pmatrix}$ can be estimated by defining a mean lattice site for each particle i in one frame at time t as the mean of all n nearest neighbor coordinates $\mathbf{x}_k(t)$:

$$\mathbf{r}(t) = \mathbf{x}_i(t) - \frac{1}{n}\sum_{i=1}^{n}\mathbf{x}_k(t) \tag{4.17}$$

where the coordinates $\mathbf{x}_i(t), \mathbf{x}_k(t)$ are with respect to the image axis. The mean lattice site is then time-dependent. Motions not related to the particle oscillation like rotations or shifts of the whole system are therefore eliminated in $\mathbf{r}_i(t)$. This method is only applicable as long as the crystal structure is well defined. If the nearest neighbors of a particle change from one frame to the next, e.g. due to diffusion, it is not possible to define the mean lattice site clearly, and large jumps in $\mathbf{r}$ can be the consequence.

The displacement and velocity distributions are given by the histograms of the respective quantity. Gaussian fits $A \propto e^{-(\xi - \bar{\xi})^2/(2\sigma_\xi^2)}$ with $\xi = r_x, r_y, v_x, v_y$ to each component of those quantities provide the widths σ_r, σ_v separately for the x- and y-direction. In an isotropic system, the fits produce the same outcome for both directions. Such fits were performed with the same Levenberg-Marquardt-Algorithm as it was introduced in Sect. 4.2.2 for the pair correlation function, giving the goodness of the fit χ^2 and the 1-σ uncertainty for the fit parameters for a qualitative evaluation.

In the case of problems with the mean lattice site identification, another method to at least obtain the dispersion of the displacement was also presented Sect. 4.2.2 through the relation $\sigma_r = \sigma_0/\sqrt{2}$ of σ_r to the dispersion of the interparticle distances (4.8). σ_0 can either be obtained from the fit to the pair correlation function, or directly from a Gaussian fit to the histogram of interparticle distances in one image.

The obtained widths σ_r, σ_v are connected to the particle temperature and Einstein frequency:

$$\sigma_r = \sqrt{\frac{k_B T}{m \Omega_E^2}}, \quad \sigma_v = \sqrt{\frac{k_B T}{m}} \tag{4.18}$$

It is therefore possible to obtain averaged quantities like the particle temperature by a simple measurement of all particle coordinates in one image. On the other hand, the same distributions can be calculated for a single particle from a long enough time series. This then yields a locally defined temperature as an average over time.

In an ergodic system, the averages over the particle ensemble are equal to those over time. Assuming that the ergodic hypothesis holds for plasma crystals, one could simply record images of a few particles out of a larger ensemble, at a spatial resolution high enough to calculate velocities. From long time series of single particle trajectories then a temperature representing that of the particle ensemble could be estimated. This way one could avoid the bad spatial resolution accompanying the recording of a huge ensemble, where velocities are subject to large uncertainties. An analysis of the ergodicity of (small) plasma crystals showed that the assumption of ergodicity might be wrong [24], it might be valid though in large enough systems, where external parameters change only slowly compared to the spatial scale of the system.

The above interpretation of the distribution functions as thermodynamic characterization of the particle system by means of a temperature is not strictly valid, because the complex plasma is an open system, not in thermal equilibrium with its surroundings. However, the continuous interaction of the particles with plasma constituents establish a certain equilibrium with regard to the particle energy (damping by collisions with neutral atoms, heating by inelastic collisions with ions). Then the system can be described by a kinetic temperature behaving similar to an equilibrium temperature as was argued in [20, 25]. In fact, experimentally obtained velocity distributions in two-dimensional complex plasmas are often found to be Gaussian distributed, defining a particle temperature from the Maxwellian model (e.g. in [4, 26, 27]). A theoretical approach to the velocity distribution yielded an Maxwellian distribution with an effective particle temperature two times larger than the temperature of the ions, following from inelastic collisions with ions in the calculations [28].

References

1. F.M. Peeters, X. Wu, Wigner crystal of a screened-Coulomb-interaction colloidal system in two dimensions. Phys. Rev. A **35**(7), 3109–3114 (1987)
2. X. Wang, A. Bhattacharjee, S. Hu, Longitudinal and transverse waves in Yukawa crystals. Phys. Rev. Lett. **86**(12), 2569–2572 (2001)
3. S. Nunomura, J. Goree, S. Hu, X. Wang, A. Bhattacharjee, Dispersion relations of longitudinal and transverse waves in two-dimensional screened Coulomb crystals. Phys. Rev. E **65**, 066402–111 (2002)
4. S. Nunomura, J. Goree, S. Hu, X. Wang, A. Bhattacharjee, K. Avinash, Phonon spectrum in a plasma crystal. Phys. Rev. Lett. **89**(3), 035001 (2002)
5. S.K. Zhdanov, S. Nunomura, D. Samsonov, G.E. Morfill, Polarization of wave modes in a two-dimensional hexagonal lattcie using a complex (dusty) plasma. Phys. Rev. E **68**, 035401 (2003)
6. S. Nunomura, S. Zhdanov, D. Samsonov, G.E. Morfill, Wave spectra in solid and liquid complex (dusty) plasmas. Phys. Rev. Lett. **94**, 045001 (2005)
7. J.R. Shewchuk. Triangle: Engineering a 2D Quality Mesh Generator and Delaunay Triangulator. In: M.C. Lin, D. Manocha (eds.), *Applied Computational Geometry: Towards Geometric Engineering,* number 1148 in Lecture Notes in Computer Science, pp. 203–222, Berlin, may 1986. First ACM Workshop on Applied Computational Geometry, Springer.
8. J.M. Burgers, Geometrical considerations concerning the structural irregularities to be assumed in a crystal. Proc. Phys. Soc. **52**, 23–33 (1940)
9. C. Kittel, Introduction to solid state physics. (Wiley, Toronto, 1976)
10. D. C. Wallace, *Statistical Physics of Crystals and Liquids*, Chap. 5. World Scientific Publishing Co. Pte. Ltd Singapore (2002)
11. D.G. Grier, C.A. Murray, The microscopic dynamics of freezing in supercooled colloidal fluids. J. Chem. Phys. **100**(12), 9088–9095 (1994)
12. R.A. Quinn, C. Cui, J. Goree, J.B. Pieper, H. Thomas, G.E. Morfill, Structural analysis of a coulomb lattice in a dusty plasma. Phys. Rev. E **53**(3), R2049–R2052 (1996)
13. B.I. Halperin, D.R. Nelson, Theory of two-dimensional melting. Phys. Rev. Lett. **41**, 121 (1978)
14. V.L. Berezinskii, Destruction of long-range order in one-dimensional and two-dimensional systems having a continuous symmetry group I. Classical systems. Sov. Phys. JETP **32**(3), 493–500 (1971)
15. V.L. Berezinskii, Destruction of long-range order in one-dimensional and two-dimensional systems possessing a continuous symmetry group II. Quantum systems. Sov. Phys. JETP **34**(3), 1144–1156 (1971)
16. C.B. Markwardt, Non-linear Least-squares Fitting in IDL with MPFIT. In: D.A. Bohlender, D. Durand, P. Dowler (eds) Astronomical Society of the Pacific Conference Series, vol 411, pp 251–254 (2009)
17. T.V. William, H.P. William, A.T. Saul, P.F. Brian, Numerical Recipes in C. Cambridge University Press, Online Edition: http://www.library.cornell.edu/nr/bookcpdf.html, 2nd edn. (2002)
18. D.R. Nelson, B.I. Halperin, Dislocation-mediated melting in two dimensions. Phys. Rev. B **19**, 2457 (1979)
19. K.J. Strandburg, J.A. Zollweg, G.V. Chester, Bond-angular order in two-dimensional Lennard-Jones and hard-disksystems. Phys. Rev. B **30**(5), 2755–2759 (1984)
20. R.A. Quinn, J. Goree, Single-particle Langevin model of particle temperature in dusty plasmas. Phys. Rev. E **61**(3), 3033–3041 (2000)
21. P.S. Epstein, On the resistance experienced by spheres in their motion through gases. Phys. Rev. **23**(6), 710–733 (1924)
22. U. Konopka, Wechselwirkungen geladener Staubteilchen in Hochfrequenzplasmen.PhD thesis, Fakultät für Physik und Astronomie der Ruhr-Universität-Bochum (2000)
23. B. Liu, J. Goree, V. Nosenko, L. Boufendi, Radiation pressure and gas drag forces on a melamine-formaldehyde microsphere in a dusty plasma. Phys. Plasma. **10**(1), 9–20 (2002)

24. C.A. Knapek, Experimental investigation of dynamical properties and ergodicity in plasma crystals. Diploma thesis, Technische Universität München (2004)
25. R.A. Quinn, J. Goree, Experimental test of two-dimensional melting through disclination unbinding. Phys. Rev. E **64**, 051404 (2001)
26. V. Nosenko, J. Goree, A. Piel, Laser method of heating monolayer dusty plasmas. Phys. Plasmas. **13**, 032106 (2006)
27. J.B. Pieper, J. Goree, Dispersion of plasma dust accoustic waves in the strong-coupling regime. Phys. Rev. Lett. **77**(15), 3137–3140 (1996)
28. A.G. Zagorodny, P.P.J.M. Schram, S.A. Trigger, Stationary velocity and charge distributions of grains in dusty plasmas. Phys. Rev. Lett. **84**(16), 3594–3597 (2000)

Chapter 5
Estimation of the Coupling Strength Γ

One of the parameters characterizing complex plasmas as it was introduced in Sect. 2.3 is the coupling parameter Γ—the ratio of the mean potential to the mean kinetic energy. For a Yukawa-type potential $Q e^{-r/\lambda_D}/(4\pi\epsilon_0 r)$ with a screening length λ_D, the particle charge Q and a particle temperature $k_B T$ this yields

$$\Gamma = \frac{\langle E_{\text{pot}} \rangle}{\langle E_{\text{kin}} \rangle} = \frac{Q^2 F(\kappa)}{4\pi\epsilon_0 \Delta k_B T} = \Gamma_{\text{eff}} F(\kappa) \qquad (5.1)$$

with the screening parameter $\kappa = \Delta/\lambda_D$. The mean kinetic energy is $k_B T/2$ per degree of freedom (in 2D this are two), and Δ is the equilibrium interparticle distance. The factor $F(\kappa)$ contains the modification of the potential due to screening and depends on the arrangement of the nearest neighbors; Γ_{eff} is the effective coupling parameter for an unscreened potential.

To obtain Γ for a specific experiment, one needs to measure particle charge and temperature, interparticle distance and the screening parameter κ. Further one needs a model for the calculation of $F(\kappa)$. Except of the interparticle distance, those quantities are not easily acquired and often subject to large uncertainties. Especially the measurement of the charge by the use of wave spectra analysis [1–3] can be subject to inaccuracies of 15%. Other methods for charge measurement like vertical particle oscillations [4] involve additional experimental effort.

Here, a method for the estimation of Γ is presented which solely depends on particle dynamics. In Fig. 5.1a the situation is sketched for three particles. The center particle with charge Q_1 sits at the mean distance Δ from its neighboring particles with charges $Q_2 = Q_3 = Q_1$. The distance is given as the position where the potential energy is minimized. The outer particles are held fixed, while the center particle has a temperature $k_B T$ and therefore can move a distance 2σ inside the potential well (solid lines) where the kinetic energy exceeds the potential energy.

In the two-dimensional case, Fig. 5.1b, the area A_1 (horizontally hatched) of the hexagon of the nearest neighbors around the center particle reflects the mean potential energy, because particles take their equilibrium positions due to their charge and the potential shape. This area is obviously proportional to the squared interparticle

C. A. Knapek, *Phase Transitions in Two-Dimensional Complex Plasmas,*   47
Springer Theses, DOI: 10.1007/978-3-642-19671-3_5,
© Springer-Verlag Berlin Heidelberg 2011

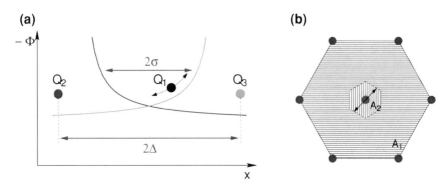

Fig. 5.1 a Potential well generated by two particles with equal charges Q_2, Q_3 as seen by the center particle with the same charge Q_1. The surrounding particles are 2Δ apart, with Δ being defined by the minimum of the potential energy. The center particle can move inside the potential well up to a distance of 2σ, depending on its kinetic energy, or temperature. **b** Areas in a two-dimensional hexagonal structure. The center particle can visit the vertically hatched area A2 defined by its temperature and the potential distribution

distance Δ^2. Again all particles have equal charge, and the outer particles are fixed. The center particle, oscillating around its mean position, visits the area A_2 (vertically hatched), defined by the confinement distance σ_r of the repulsive potential well generated by the surrounding particles. The driving force for this motion defining the range σ_r is the particle temperature. σ_r is the displacement dispersion introduced in Sect. 4.3, and A_2 is proportional to its squared value.

Then the ratio of the areas A_1 and A_2 should represent the ratio of mean potential to mean kinetic energy:

$$\frac{A_1}{A_2} = \frac{\Delta^2}{\sigma_r^2} = \tilde{\Gamma} \tag{5.2}$$

The first equality holds under the assumption of equally shaped areas.

$\tilde{\Gamma}$ is a quantity which can be measured directly from the particle coordinates, provided the spatial resolution is high enough to resolve the displacement from the mean lattice site.

For this purpose, the particle displacements $\mathbf{r}$ and Δ will be calculated as localized quantities. A local coordinate system is introduced for each particle separately. Its center is the mean lattice site defined as the time dependent mean position of all n nearest neighbors of particle i. The time dependence removes systemic trends like rigid-body drifts of the system. In the local coordinate system, for each particle i a localized, time dependent displacement $\mathbf{r}_i(t)$, nearest neighbor distance $\Delta_i(t)$ and velocity $\mathbf{v}_i(t)$ can be calculated:

$$\mathbf{r}_i(t) = \begin{pmatrix} r_{x,i}(t) \\ r_{y,i}(t) \end{pmatrix} = \mathbf{R}_i(t) - \frac{1}{n}\sum_{j=1}^{n} \mathbf{R}_j(t) \tag{5.3}$$

$$\Delta_i(t) = \frac{1}{n} \sum_{j=1}^{n} |\mathbf{R}_i(t) - \mathbf{R}_j(t)| \tag{5.4}$$

$$\mathbf{v}_i(t) = [\mathbf{r}_i(t + \delta t) - \mathbf{r}_i(t)]/\delta t \tag{5.5}$$

The summation is over the n nearest neighbors found by performing Delauney triangulations. $\mathbf{R}_k(t) = \begin{pmatrix} x_k(t) \\ y_k(t) \end{pmatrix}$ are the original particle coordinates as to the image axis found by the tracking algorithm. From here on, indices x and y will always refer to the direction x, y of the image axis, and $r_{x,y}$ and $v_{x,y}$ mean the quantities in the local coordinate system with axis parallel to the image axis. Coordinates (x, y) refer to the coordinates in the image coordinate system.

The following sections present the results of the first experiments performed to estimate the coupling strength $\tilde{\Gamma}$ using the method proposed above; they were also published in [5].

The experimental parameters and the uncertainties in the coordinates and displacements are given in Sects. 5.1 and 5.2. The localized particle dynamics are presented and analyzed in Sect. 5.3. The underlying theoretical model will be introduced in Sect. 5.4. The last Sect. 5.5 summarizes the results.

5.1 Experimental Parameters

The basic setup of the plasma chamber and discharge generation has been described in Sect. 3.1. For the estimation of $\tilde{\Gamma}$ an undisturbed two-dimensional crystalline system was needed.

For this purpose, a two-dimensional crystal, consisting of melamine-formaldehyde spheres with a diameter of $9.19 \pm 0.09\,\mu\text{m}$, was generated in an argon discharge with a peak-to-peak voltage of $-172\,\text{V}_{PP}$ and $1.94\,\text{Pa}$ neutral gas pressure. The particles were levitated at a height of $6.4\,\text{mm}$ above the lower electrode. The crystal dimensions were $\approx 41.9 \times 17.4\,\text{mm}^2$. It was verified that no particles were located above or below the layer by scanning vertically through the chamber and searching for particles at another height. The crystal was confined from two sides by parallel tungsten wires $59.2\,\text{mm}$ apart and a few millimeters above the crystal plane. The wires were at a negative floating potential caused by electrons collected from the plasma.

Two recordings, numbered I and II, were done at a high spatial and temporal resolution. They show different sections of the same crystal. A $105\,\text{mm}$ lens plus three additional distance-rings with a total length of $49.5\,\text{mm}$ provided a spatial resolution of $6.74 \times 10^{-3}\,\text{mm/px}$. The field of view (fov) had a size of $6.9\,\text{mm} \times 6.9\,\text{mm}$. The total recording length for each experiment was 6144 frames at a frame rate of 500 fps.

In experiment I approximately 111 particles where located in the field of view including a 5-fold and a 7-fold defect forming a dislocation, while in experiment II a perfect crystal with approximately 115 particles was recorded at a higher

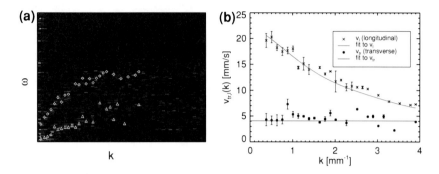

Fig. 5.2 a Wave spectrum ω vs. k. Marked in *yellow* are positions identified as the longitudinal (*open diamond*) and transverse branch (*open triangle*). **b** Transverse (*filled circle*) and longitudinal (*multiplication symbol*) sound velocities obtained from the wave spectrum in the *left panel*, plotted vs. wave number k. *Red* and *blue lines* are fits according to theory (Sect. 4.1). The fits yield the sound speed c_0 and therefore the particle charge from the transversal (*blue line*), and κ from the longitudinal velocity (*red line*)

illumination level of the laser. The approximate values come from the fact that particles at the image edges sometimes vanish out of the field of view during a recording.

An additional set of images, referred to as experiment III, of the same crystal at a lower spatial resolution of 0.0325 mm/px was recorded, with a fov size of 33.28 mm × 33.28 mm. Pressure and discharge parameter settings were kept unchanged. It contained 2248 particles in the field of view, providing a much larger system. This allowed a wave spectra analysis as described in Sect. 4.1 to estimate the particle charge Q and screening parameter κ. The wave spectrum and sound velocities are shown in Fig. 5.2. The particle charge is estimated to be $Q = -10500 \pm 100$ elementary charges, and $\kappa = 0.86 \pm 0.37$.

Charge and κ—and also the particle temperatures $k_B T$, which will be derived from the velocity distributions—will be needed in the later Sect. 5.4 to compare the theoretical model with the assumption that the coupling parameter is given by $\tilde{\Gamma}$.

Table 5.1 lists all experimental parameters and additionally some properties of the crystal.

With the gas parameters and particle diameter (4.13) yields the damping rate due to collisions with neutral gas atoms $\nu_{Ep} = 2.28$ Hz according to the reflection index $\delta = 1.26$ measured in [6], or $\nu_{Ep} = 2.68$ Hz with $\delta = 1.48$ from [4], for δ measured from horizontal oscillations.

Figure 5.3a shows an example of the images for the high resolution data of experiment I. For better visibility, 100 consecutive frames were overlaid to enhance the particle image size and brightness. Figure 5.3b shows the Voronoi cells of one frame with the defect lattice sites marked by vertical hatching (5-fold) and horizontal hatching (7-fold).

Table 5.1 Experimental parameters and basic properties of the plasma crystal for the experiments used to measure $\tilde{\Gamma}$

	I	II	III
	Discharge settings		
rf power forward/reflected		10/0 W	
Self bias		$-69 \ldots -70$ V	
Peak-to-peak voltage		-172 V	
Gas		Ar	
Flow rate		4.5 sccm	
Neutral gas pressure	1.946 Pa		1.947 Pa
	Particles		
Particles		MeF, $\varnothing$ 9.19 $\pm$ 0.09 μm	
Particle mass density		1.51 g/cm^3	
Epstein drag coefficient		2.28 Hz ($\delta = 1.26$), 2.68 Hz ($\delta = 1.48$)	
Particle charge		-10500 ± 100 e	
Screening parameter		0.86 $\pm$ 0.37	
Height of particles above the electrode	6.4 mm		6.7 mm
	Recording		
Recorded frames		6144	
Frame rate		500 fps	
Resolution	6.74 $\times$ 10^{-3} mm/px		0.0325 mm/px
Aperture		2.8	5.5
Illuminating laser power	88 mW	106 mW	200 mW

The Epstein drag coefficient depends strongly on the reflection index δ used in its calculation, therefore two extreme values are given according to the quantity of $\delta = 1.26$ and $\delta = 1.48$ from the Refs. [6, 4]

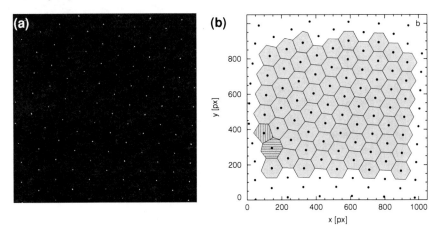

Fig. 5.3 a 100 consecutive images of experiment I, averaged for better visibility. **b** Voronoi cells of one image of experiment I. The *vertically* and *horizontally* hatched cells mark a 5- and a 7-fold defect, respectively

Table 5.2 Pixels per particle, pixel-noise and errors of absolute coordinates for experiments I, II and III for frame rates of 500 fps and 500/3 fps

	I	II	III
	500 frames per second		
Number of pixels/particle	6.7	12	6.7
Pixel-noise level	8.4	11.7	12.6
Error absolute coordinates $\delta_{r,\text{abs}}$ [px]	0.074	0.072	0.082
	500/3 frames per second		
Pixel-noise level	4.9	6.8	7.3
Error absolute coordinates $\delta_{r,\text{abs}}$ [px]	0.070	0.054	0.073

5.2 Uncertainties

The quantity of the error in the coordinates is identified by comparing the number of illuminated pixels per particle and the pixel-noise level with the outcome of the simulations in Chap. 9.

The number of pixels per particle is obtained during the particle tracking and then averaged over all particles. In experiment I an average of 6.7 pixels per particle was found. Experiment II was recorded at a higher illumination level. The particle images are larger, on average consisting of 12 pixels/particle. The experiment III with a low spatial resolution had mean particle sizes of 6.7 pixels/particle.

The pixel-noise level was estimated using the method of intensity fluctuations in the pixels composing the particle images, as mentioned earlier in Sect. 3.3 and explained in Chap. 9. Noise levels of 8.4, 11.7 and 12.6 (in units of intensity with a maximum of 255) for experiments I, II and III respectively, were found. The higher noise levels for the data sets II and III are attributed to the corresponding higher illumination laser power of 106 and 200 mW compared to 88 mW in experiment I.

For noise-levels >4 and particles >6 pixels the combined error $\delta_{r,\text{abs}}$ in absolute coordinates caused by tracking and pixel-noise effects was found to be well approximated by a statistical error.

Table 5.2 lists the particle sizes, noise levels and the range of respective standard deviations of the error distributions of the absolute particle positions for the cases of interest here. Also shown are the expected errors for a frame rate of $500/3 \approx 166$ fps. To lower the frame rate and improve the shape of the particles in the images, each three consecutive images were averaged as to their intensity. This reduces the pixel-noise by a factor of $1/\sqrt{3}$. The number of pixels per particle was practically not affected by this procedure. The necessity for this procedure becomes evident from the quantity of the error for particle displacements at the full frame rate.

For distances between two coordinates, e.g. velocities and displacements, the error is statistical all the time, but depends on the real distance between the two positions. It is presented graphical in Fig. 5.4 for experiments I–III. The figure shows the width of the error distribution plotted vs. the width of the real displacement distribution, both

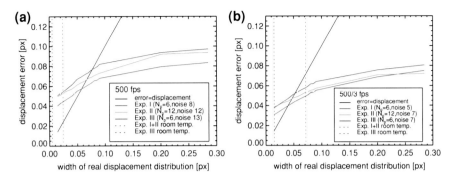

Fig. 5.4 Errors of particle displacements vs. real displacement for particle image sizes and noise-levels comparable to the experiment I, II, III. The data are taken from simulations of artificial particles. The *dotted lines* show the width of a velocity distribution of particles at room temperature at the respective spatial and temporal resolutions. The error must be smaller than the bisecting line for valid measurements. Errors for: **a** Pixel-noise values from the original images at 500 fps. **b** Pixel-noise values for images averaged over three consecutive frames, leading to a noise level reduced by $1/\sqrt{3}$

converted to units of pixels. The data come from the artificial particle error analysis with parameters meeting the conditions of the experiments. The real displacement means here the distance between any two positions. For velocities, this distance has to be divided by the time step.

The solid black line is the bisecting line where the error equals the real displacement. If one wants to measure the width of the distribution of real values, the error must be below this line, i.e. the error must be smaller than the value which is to be measured. For particles at room temperature the dotted vertical lines show the quantity of the width of the velocity distribution in units of pixels for the spatial and temporal resolutions of experiments I–III. It is clear from Fig. 5.4a, that at 500 fps it is not possible to obtain a valid temperature from the velocity distribution for particles at room temperature. For 500/3 fps, Fig. 5.4b, the error is smaller than the real value for the high spatial resolution of experiments I and II. Since experiment III will be used only for wave spectra analysis, the high temporal resolution is not essential, and an even lower frame rate of 83 fps is applicable. For the measurement of σ_r the frame rate is not important. However it was found that σ_r is much larger ($\approx 10\times$) than the errors plotted in Fig. 5.4, so it should not be affected much by the uncertainties.

5.3 Localized Particle Dynamics

Figure 5.5 shows all particle trajectories of experiment I (a) and II (b) for the whole time series of overlaid images with the reduced frame rate 500/3 fps. Colored trajectories where analyzed. The dislocation consisting of a 5- and a 7-fold defect

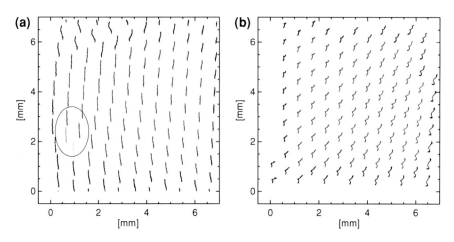

Fig. 5.5 Particle trajectories for the whole measurement time. All colored trajectories where ana-lyzed. The colors mean 5-fold (*green*), 6-fold (*blue* and red) and 7-fold (*yellow*) lattice sites. The blue trajectories will be referred to later. Trajectories colored black are edge particles not used in the analysis. **a** Experiment I. The two *red particles* enclosed by the *black ellipse* showed a different behavior from all other 6-fold particles and will be referenced later. **b** Experiment II

in Fig. 5.5a is marked green and yellow. All other particles (red and blue) have hexagonal unit cells. The trajectories colored blue in experiment I are close to a dis-turbance visible in the upper left line of particles and will be referred to later. Due to out-of-view neighbor particles it is not possible to define the mean lattice site for particles at the image edges (black) which are therefore not considered.

The 5-fold defect particle in experiment I vanishes and re-emerges several times thus fragmenting its trajectory. The particle is badly illuminated, presumably due to a small vertical offset from the crystal plane, and at some times a further decrease of its intensity, maybe due to small additional vertical oscillations, prevents tracking. This affects the velocity calculation as well as the estimation of the mean lattice site of its nearest neighbors. To prevent jumps in the mean lattice sites, only those frames were considered for the 5-fold and its neighbors in which all particles were visible.

5.3.1 Local Coordinate System

For each time-series, the histograms of displacement components $r_{x,i}(t)$, $r_{y,i}(t)$ and velocity components $v_{x,i}(t)$, $v_{y,i}(t)$, calculated in the local coordinate system intro-duced in (5.5), are obtained separately. The histograms were Gaussian distributed, and are interpreted in context of the statistical evaluation from Sect. 4.3. The inter-particle distance Δ is calculated in each frame locally for each nearest neighbor to a particle i and then averaged over all neighbors, yielding a time-series of $\Delta_i(t)$ which appears Gaussian distributed around a local mean Δ. The Gaussian fits to the histograms produced the widths $\sigma_{v,x}$, $\sigma_{v,y}$, $\sigma_{r,x}$, $\sigma_{r,y}$ and the local mean Δ for each particle.

The local particle temperature T_i for the ith particle is then given by $\sigma_{v,i}^2 = (\sigma_{v_x,i}^2 + \sigma_{v_y,i}^2)/2$. The displacement dispersion is $\sigma_{r,i}^2 = (\sigma_{r_x,i}^2 + \sigma_{r_y,i}^2)/2 \equiv \Delta_i^2/\tilde{\Gamma}_i$. With the knowledge of the Δ_i, $\tilde{\Gamma}_i$ can be calculated as a local quantity using (5.13) as $\tilde{\Gamma}_i = \Delta_i^2/\sigma_{r,i}^2$.

Particle temperature T_i, displacement dispersion $\sigma_{r,i}$, Δ_i and $\tilde{\Gamma}$ are displayed in Figs. 5.6 and 5.7 in dependence on the average mean lattice site position on the x-axis of the images for experiments I and II. The average interparticle separation obtained from the pair correlation function for the whole crystal is indicated as a solid line in Figs. 5.6a and 5.7a for comparison with the localized quantity.

Experiment I The values for particles marked green, yellow and blue in Fig. 5.5a are colored in the same manner in Fig. 5.6. Colored in red are the two particles included in the black ellipse in Fig. 5.6. Those particles are direct neighbors to the dislocation and are affected by the non continuous trajectory of the 5-fold particle.

The defect pair (green, yellow) stands out with a smaller resp. larger local interparticle distance for the 5- and the 7-fold (Fig. 5.6a). This is reflected in $\tilde{\Gamma}$ in Fig. 5.6d. The temperatures for both defects are nearly half of that of the other particles (Fig. 5.6c). This is caused by their non-continuous trajectories and the resulting small number of velocities which introduces large errors into the statistical analysis. Also the two direct neighbors of the defects marked in red have huge error bars in $k_B T$ due to the lack of good statistics. σ_r for the dislocation particles and the neighbors is in the range of the results for the other 6-folds.

Standing out are the blue trajectories with high σ_r and therefore low $\tilde{\Gamma}$. On the first look, these particles, located next to a small disturbance in the crystal in the upper line of edge particles, seem to be more mobile than the rest, though their temperature is not affected. The effect seen in the displacements is artificial, as shown in Fig. 5.6e, f. Here, the displacement was calculated by subtracting a trajectory smoothed over 100 points as the mean lattice site from the coordinates, instead of introducing the local coordinate system. In that configuration, σ_r and $\tilde{\Gamma}$ of the blue particles do not deviate from that of the other 6-folds, but now the defect particles and their closest neighbors show higher mobility.

Obviously the local coordinate system introduces large errors as soon as the neighboring particles do not behave completely linear. This is illustrated in Fig. 5.8 by examples of the displacement and velocity histograms and the particle trajectory in the localized coordinate system, compared with those where the mean motion was subtracted as a smoothed trajectory. The top line (a, b) shows histograms of a particle with six neighbors out of the bulk of the crystal, and the particle trajectory in the local coordinate system in c. The middle line (d–f) belongs to one of the blue colored particles, featuring a double-peak in the displacement distribution (d). The velocity histogram (e) is not affected and practically looks identical to that in Fig. 5.8b. The bottom line (g–i) shows the histograms for the same particle as the middle line, now not in the local coordinate system, but with a smoothed trajectory subtracted. The double-peak is gone in the displacement distribution, and comparing the particle

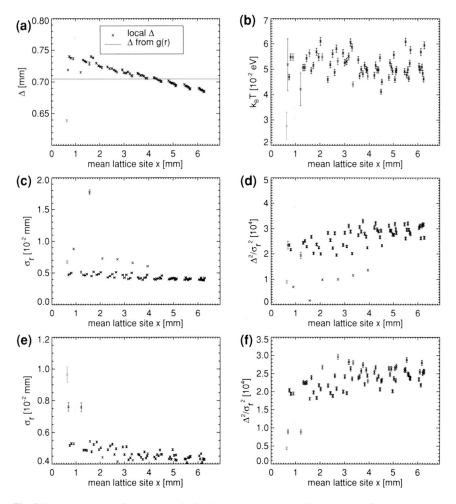

Fig. 5.6 Experiment I. Quantities calculated in the local coordinate system from histograms and plotted vs. the average mean lattice site position on the x-axis of the images. The error bars are the 1-σ uncertainties provided by the fit of Gaussian functions to the histograms. Colors are explained in the text. **a** Interparticle distance Δ, **b** particle temperature $k_B T$, **c** displacement dispersion σ_r, **d** $\tilde{\Gamma} = \Delta^2/\sigma_r^2$. The following quantities are calculated not in the local coordinate system, but by substraction of a smoothed trajectory from the original data: **e** displacement dispersion σ_r, **f** $\tilde{\Gamma}$

trajectories f, i one can see that the disturbance in the upper line of particles leads to a large deviation of the mean lattice site to the actual average particle position: in Fig. 5.8f the trajectory is off-center in the plot, and even seems to shift, therefore the elongated shape.

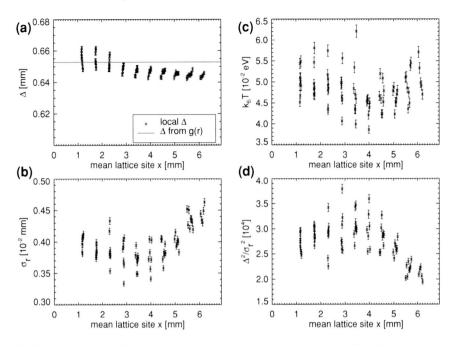

Fig. 5.7 Experiment II. Quantities calculated in the local coordinate system from histograms and plotted vs. the average mean lattice site position on the x-axis of the images. The error bars are the 1-σ uncertainties provided by the fit of Gaussian functions to the histograms. **a** Interparticle distance Δ, **b** displacement dispersion σ_r, **c** particle temperature $k_B T$, **d** $\tilde{\Gamma} = \Delta^2/\sigma_r^2$

This effect is artificial, since obviously by looking at the trajectories in Fig. 5.5, the disturbance only shifts the mean lattice site geometrically, but does not affect the trajectories of the blue colored neighboring particles.

Experiment II In the case of the almost perfect crystal of experiment II, a dependence on the x-coordinate appears in all quantities, see Fig. 5.7. Especially the temperature shows the same features in dependence on x as σ_r and $\tilde{\Gamma}$, as it would be expected, since $\tilde{\Gamma} \propto 1/T \propto 1/\sigma_r^2$.

The next section emphasizes the local characteristics of the individual particles. The influence of the location within the crystal on the particle dynamics is evaluated further.

5.3.2 2D Maps

Coordinate maps of the particle system were generated containing all information obtained in the former chapter. At each location (x, y) of a particle the Voronoi cell is colored according to the time-averaged localized data. This is done for the temperature T_i, the interparticle distance Δ_i, the width $\sigma_{r,i}$ and $\tilde{\Gamma}_i$.

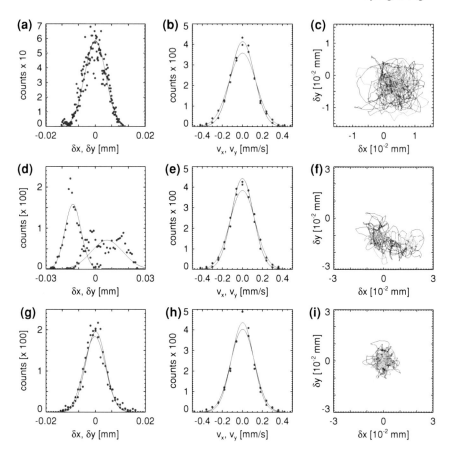

Fig. 5.8 Examples of displacement (*left*) and velocity (*middle*) histograms with Gaussian fits, and local particle coordinates (*right*). The components x, y and v_x, v_y were examined separately and are colored *red* (x, v_x) and *blue* (y, v_y). **a–c** 6-fold lattice site; **d–f** 6-fold lattice site close to a small disturbance, quantities calculated in the local coordinate system; **g–i** the same 6-fold lattice site, quantities calculated by taking the smoothed particle trajectory as the mean lattice site

To identify the role of structural properties of the crystal, the bond order parameter Ψ_6 introduced in Sect. 4.2.4 has also been calculated at each particle position as the average over time of $\Psi_6(t)$.

Experiment I Figure 5.9 presents the results for experiment I. The dislocation is clearly identified at the lower left corner by its deviating values in all maps. Disregarding the defects, the Δ-map (Fig. 5.9a) reveals a density gradient 0.68–0.75 mm, from right 0.68 mm to left 0.75 mm, which is caused by the weak horizontal compression produced by the floating wires located to the right and left. Obviously, the field of view was not centered between the wires.

The distribution of T (Fig. 5.9b) across the crystal is random within a range of 0.041–0.061 eV, slightly above room temperature, for all particles except the defect

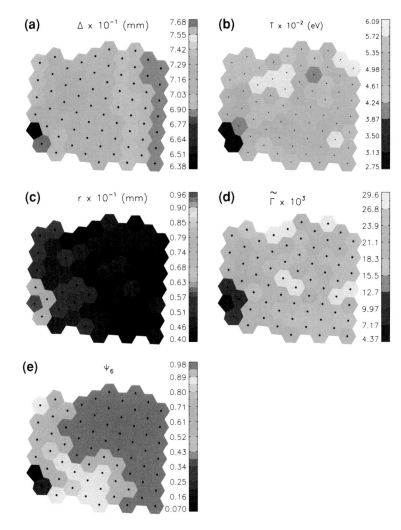

Fig. 5.9 Experiment I: 2D maps of local crystal parameters. The Voronoi cell around each particle is color coded according to the value of the particular measured quantity (see *colorbars* on the *right side* of each picture). The displacement dispersion and coupling parameter were calculated by substraction of the smoothed particle trajectory from the original trajectory. The (compressing) wires were arranged along the vertical image axis to the left and right of the plasma crystal. **a** Interparticle distance $\mathbf{\Delta}$, **b** Particle temperature $k_B T$, **c** Displacement dispersion σ_r, **d** Coupling parameter $\tilde{\Gamma}$, **e** Bond order parameter Ψ_6

lattice sites, which deviate strongly in T due to the poor statistics mentioned above. Remarkably, neither the defects nor the aforementioned disturbance at the upper part of the crystal influence the particle temperature in their vicinity.

Figure 5.9c, d show the maps of σ_r and $\tilde{\Gamma}$. Due to the problems of the localized coordinate system mentioned if the last chapter, the presented values are those obtained from subtracting the smoothed particle trajectory in the displacement calculation. While σ_r lies mostly in the range of 0.0040–0.0055 mm, the 5- and 7-fold and two of their nearest neighbors stand out with a much higher displacement of 0.0075–0.0096 mm. This is reflected in a low $\tilde{\Gamma} < 10000$ (Fig. 5.9d) while the influence of the gradient of Δ on $\tilde{\Gamma}$ is not noticeable. Elsewhere, $\tilde{\Gamma}$ falls in the range 15000–30000.

The map of the local bond order parameter is plotted in Fig. 5.9e. The defect pair exhibits a very low order, as it is expected, since Ψ_6 is defined for a hexagonal cell. The hexagonal bond order also reacts sensitively in the wider vicinity of the defect. Also the gradient in the interparticle distance might have some influence on Ψ_6, since in average it decreases from 0.98 at the right side (smaller Δ) to 0.89 on the left side (larger Δ).

Experiment II Another part of the crystal was recorded in experiment II without any defects or disturbances within the field of view. The results, now calculated in the local coordinate system, are shown in the maps in Fig. 5.10. The gradient in Δ (Fig. 5.10a) is clearly correlated to the gradient in Ψ_6 (Fig. 5.10e). The temperature (Fig. 5.10b) is distributed across the crystal in a range of 0.038–0.062 eV, with smaller values accumulated in the center region, as it was already indicated in the x-dependence in Fig. 5.7. Accordingly, smaller values of σ_r and higher values of $\tilde{\Gamma}$ appear in that region (Fig. 5.10c, d).

Conclusion The localized coordinate system works well for stationary crystal structures without lattice defects. If non-stationary processes take place, the mean lattice sites are not identified correctly, giving misleading results. In that case, the particle displacements have to be calculated by other methods, as it was done for experiment I. Defect sites are then easily identified by significantly decreasing $\tilde{\Gamma}$ even in the closer neighborhood of the defect itself. Further, $\tilde{\Gamma}$ is not visibly affected by small linear density gradients. The essential factor influencing $\tilde{\Gamma}$ locally is the average displacement of a particle from its mean lattice site.

The particle temperature in both experiments scatter around 0.05 eV. This is twice the room temperature of $\approx 20°C$, corresponding to 0.025 eV. The spatial and temporal resolution during the experiment were high enough to measure the real particle temperatures from the velocity distributions, considering both the required time scales and the uncertainties. The time scale defined by the oscillation frequency of the particles within their nearest neighbor cells is of the order of 0.03–0.04 s (this is the inverse Einstein frequency which will be derived in the next chapter and shown in Table 5.3). The time step between frames with the reduced frame rate 166 fps is sufficiently small with 0.006 s. The measured temperature is therefore real, and particles actually are above room temperature. Assuming that the temperature of the neutral gas, and that of the ions is at room temperature (see Chap. 2 and basics of complex plasmas), the measured particle temperature is consistent with the calculations by Zagorondy in [7], who finds an effective particle temperature twice the temperature of the ions. The reason stated there is the additional heating of dust particles by inelastic charging collisions with ions, in contrast to pure Brownian motion.

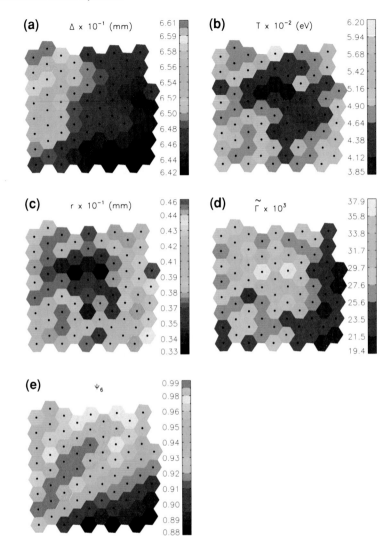

Fig. 5.10 Experiment II: 2D maps of local crystal parameters. The Voronoi cell around each particle is color coded according to the value of the particular measured quantity (see *color-bars* on the *right side* of each picture). The displacement dispersion and coupling parameter were calculated in the local coordinate system. The (compressing) wires were arranged along the vertical image axis to the left and right of the plasma crystal **a** Interparticle distance Δ, **b** Particle temperature $k_B T$, **c** Displacement dispersion σ_r, **d** Coupling parameter $\tilde{\Gamma}$, **e** Bond order parameter Ψ_6

Table 5.3 Listing of some quantities calculated by different methods

	Experiment I	Experiment II
Charge and κ from wave spectra analysis		
Q_{ws}	$-10500 \pm 100\,e$	
κ_{ws}	0.86 ± 0.37	
$f(\kappa)$ from theory, using κ_{ws}		
$f(\kappa)$	2.23 ± 0.09	
$f_{NN}(\kappa)$	1.82 ± 0.02	
Δ, $k_B T$ calculated from data		
$\langle \Delta \rangle$	0.710 ± 0.017 mm	0.649 ± 0.005 mm
$\langle k_B T \rangle$	0.051 ± 0.005 eV	0.048 ± 0.004 eV
Ω_0, Ω_E, $\Omega_{E,NN}$ from wave spectra charge		
$\Omega_0 = Q_{ws}/\sqrt{4\pi\epsilon_0 m \Delta^3}$	10.8 ± 0.1	12.3 ± 0.1
$\Omega_E = \Omega_0 f(\kappa)$	24.0 ± 1.0	27.4 ± 1.2
$\Omega_{E,NN} = \Omega_0 f_{NN}(\kappa)$	19.5 ± 0.3	22.3 ± 0.3
Γ from wave spectra charge		
$\Gamma = \Gamma_{\text{eff}} f(\kappa)^2$	21000 ± 3000	25000 ± 3000
$\Gamma_{NN} = \Gamma_{\text{eff}} f_{NN}(\kappa)^2$	14000 ± 1500	17000 ± 1600
Ω_E, $\tilde{\Gamma}$ from particle motion		
$\Omega_E = \sigma_v/\sigma_r$	24 ± 4	28 ± 2
$\tilde{\Gamma} = \Delta^2/\sigma_r^2$	23000 ± 4000	27000 ± 4000
Charge from $\tilde{\Gamma}$ and $f(\kappa)$		
Q	$-10800 \pm 1200\,e$	$-11000 \pm 1000\,e$
Q_{NN}	$-13300 \pm 1400\,e$	$-13500 \pm 1100\,e$

Q_{ws} and κ_{ws} are the charge and screening parameter obtained from wave spectra analysis. $f(\kappa)$, $f_{NN}(\kappa)$ are calculated from the theory with κ_{ws}. The index NN corresponds to the nearest neighbor approximation in (5.12). $\langle \Delta \rangle$ and $\langle k_B T \rangle$ are averages over all particles and frames. With these values follow the frequencies Ω_0, Ω_E and $\Gamma = f^2(\kappa)\Gamma_{\text{eff}}$. The Einstein frequency is also calculated independent from the charge as σ_v/σ_r and $\Gamma = \Delta^2/\sigma_r^2$. From this follow the particle charges in the last two lines

5.4 Theory for the Estimation of $f(\kappa)$

In this chapter, the theory for the calculation of $f(\kappa) = \Omega_E/\Omega_0$ will be explained. It will be shown, that the factor $F(\kappa)$, introduced in the beginning of Chap. 5 in the equation for the coupling parameter $\Gamma = F(\kappa)\Gamma_{\text{eff}}$, is related to this quantity. It describes the influence of the screening on the potential energy. It will also be shown that the quantity $\tilde{\Gamma} = \Delta^2/\sigma_r^2$, which was calculated from the particle coordinates in a two-dimensional crystalline system in the last chapter, is equal to Γ [5].

In the plasma crystal, particles oscillate about their equilibrium positions within the lattice with the plasma crystal Einstein frequency Ω_E [8]. This frequency depends on the Yukawa-type interparticle potential, and this in turn depends on the particle charge.

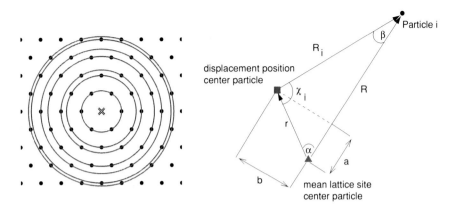

Fig. 5.11 *Left* Consecutive shells (*solid lines*) around a test particle (center **x**). The first shell contains the nearest neighbors to the center particle, the next shell includes the next set of particles, and so on. *Right* Construction to obtain the distance R_i in (5.7) between the displacement position of the center particle (*filled square*) to a neighboring particle i (*filled circle*). The displacement refers to the mean lattice site (*filled triangle*)

Assuming a regular lattice of particles with the charge Q, one can choose an arbitrary test particle and divide the lattice around that particle into concentric circles in 2D, or spheres in 3D, for simplicity referred to as shells. Each shell contains a set of particles, as indicated in the left part of Fig. 5.11 for the two-dimensional case.

If the center particle has a small displacement **r** from its mean lattice site while all other particles are fixed, the potential energy $W(\mathbf{r})$ is

$$W(\mathbf{r}) = \sum_{\text{shells}} \sum_{i} Q\Phi(R_i) \tag{5.6}$$

with an isotropic interaction potential Φ and the distance R_i between the displaced center particle and the ith particle of one shell. The first sum goes over all shells.

The distance R_i can be calculated using the construction in the right part of Fig. 5.11. R is the shell radius and **r** points from the mean lattice site (filled triangle in the figure) to the displaced (actual) position of the center particle (filled square). χ_i is the angle between the displacement vector **r** and the vector $\mathbf{R}_i$ from the displaced position to any particle i. With the annotations in the figure follows:

$$\left.\begin{array}{l} \cos\alpha = a/r \\ \sin\alpha = b/r \\ (R-a)^2 + b^2 = R_i^2 \\ \alpha + \beta + \chi_i = \pi \end{array}\right\} \Rightarrow \begin{array}{l} R_i^2 = R^2 - 2Rr\cos\alpha + r^2\cos^2\alpha + r^2\sin^2\alpha \\ \qquad = R^2 - 2Rr\cos\alpha + r^2 \end{array}$$

$\cos\alpha$ can be written as $\cos\alpha = \cos(\pi - \beta - \chi_i) = -\cos\beta\cos\chi_i + \sin\beta\sin\chi_i$. For small displacements r the angle β becomes very small and the approximation $\sin\beta \approx 0$ and $\cos\beta \approx 1$ is applicable, therefore:

$$\cos \alpha \approx -\cos \chi_i$$

$$R_i^2 = R^2 + r^2 + 2Rr \cos \chi_i \tag{5.7}$$

The potential energy can be expanded around the equilibrium position $r = 0$ in a Taylor expansion, where r is just the distance from the center and the angle dependence is included in R_i:

$$W(r) = W_0 + \left.\frac{dW}{dr}\right|_{r=0} r + \frac{1}{2} \left.\frac{d^2W}{dr^2}\right|_{r=0} r^2 + O(r^3) \tag{5.8}$$

W_0 is simply a constant factor equals the minimum potential energy at $r = 0$. The first derivative $\left.\frac{dW}{dr}\right|_{r=0} = -F = 0$ at the minimum $r = 0$ is zero, there is no force acting in equilibrium. With the Yukawa interaction potential between the particles, $\Phi(R) = Qe^{-K}/(4\pi\epsilon_0 R)$ with $K = R/\lambda_D$, the Taylor expansions gives:

$$W(r) = W_0 + \frac{1}{2} \frac{Q^2}{4\pi\epsilon_0\lambda_D^3} \sum_{\text{shells}} \sum_i \frac{e^{-K}}{K} \left[(1 + 3K^{-1} + 3K^{-2}) \cos^2 \chi_i - K^{-1} - K^{-2}\right] r^2 \tag{5.9}$$

This describes a harmonic oscillator, and the acting forces for small displacements can be expressed as $F = -kr$ with a spring constant $k = m\Omega_E^2$, where m is the particle mass and Ω_E the Einstein frequency. With $d^2W/dr^2 = -dF/dr = k = m\Omega_E^2$, an expression for the Einstein frequency is obtained from (5.9):

$$\Omega_E^2 = \frac{Q^2}{4\pi\epsilon_0 m\lambda_D^3} \sum_{\text{shells}} \sum_i \frac{e^{-K}}{K} \left[(1+3K^{-1}+3K^{-2}) \cos^2 \chi_i - K^{-1} - K^{-2}\right] \tag{5.10}$$

In two-dimensional systems, $\cos \chi_i = \cos(\phi - \phi_i)$ with the angles ϕ, ϕ_i of the polar coordinates of the respective particles.

With the frequency of the one-component plasma $\Omega_0 = \sqrt{Q^2/(4\pi\epsilon_0 m\Delta^3)}$, the ratio $\Omega_E/\Omega_0 \equiv f(\kappa)$ can be calculated as:

$$f(\kappa) \equiv \frac{\Omega_E}{\Omega_0} = \sqrt{\left(\frac{\Delta}{\lambda_D}\right)^3 \sum_{\text{shells}} \sum_i \frac{e^{-K}}{K} \left[(1 + 3K^{-1} + 3K^{-2}) \cos^2 \chi_i - K^{-1} - K^{-2}\right]} \tag{5.11}$$

Considering only the first nearest neighbor shell, $R = \Delta$ and $K = R/\lambda_D = \Delta/\lambda_D \equiv \kappa$ becomes the screening parameter. For different geometries (5.11) then simplifies to

$$
\left.\begin{array}{lll}
\text{1D chain} & \text{(2 NN)} \\
\text{2D hexagonal} & \text{(6 NN)} \\
\text{3D bcc} & \text{(8 NN)} \\
\text{3D fcc, hcp} & \text{(12 NN)}
\end{array}\right\} \quad f_{NN}(\kappa) = \frac{\Omega_{NN}}{\Omega_0} = \left\{\begin{array}{l}
\sqrt{4e^{-\kappa}(1 + \kappa + \frac{1}{2}\kappa^2)} \\
\sqrt{3e^{-\kappa}(1 + \kappa + \kappa^2)} \\
\sqrt{\frac{8}{3}e^{-\kappa}\kappa^2} \\
\sqrt{4e^{-\kappa}\kappa^2}
\end{array}\right. \tag{5.12}
$$

with the number NN of particles in the first shell in parenthesis.

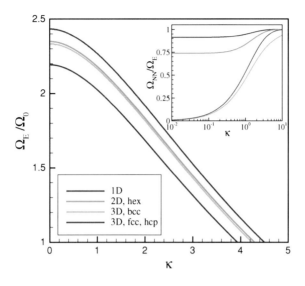

Fig. 5.12 Ratio of the Plasma Crystal Einstein (PCE) frequency Ω_E to the one-component plasma frequency Ω_0 vs. the screening parameter $\kappa = \Delta/\lambda_D$ for a 1D chain (*red*), 2D hexagonal lattice (*green*) and 3D bcc (*turquoise*) and fcc/hcp (*blue*) lattices. The inset shows the ratio of Ω_{NN} obtained from the nearest neighbor approximation in (5.12) to Ω_E in dependence of κ (figure from [5])

Figure 5.12 shows $f(\kappa)$ for different lattices in two and three dimensions. The green line, corresponding to the two-dimensional hexagonal lattice, can be fitted by the polynomial $(a + b\kappa^2)^{-1}$ with $a = 0.427$ and $b = 0.0295$. Remarkably the ratio is the same for the two-dimensional hexagonal crystal and the three-dimensional bcc lattice. The inset shows the ratio of Ω_{NN} from the nearest neighbor approximation to Ω_E. One can see that for three dimensions (blue and turquoise lines) the inclusion of more shells becomes important even for large values of $\kappa \gg 1$, while in one- and two-dimensional systems the influence of distant particles is not so large. The reason in the three-dimensional cases is the small distance between consecutive shells (in the bcc lattice for example the distance between the first and second shell is only $(2/\sqrt(3) - 1)\Delta \approx 0.15\Delta$).

The described theory holds for sufficiently small oscillations of the particles, so that there are no anharmonic effects and the linear harmonic oscillations of particles are independent from each other.

With the relation $\sigma_r^2 = k_B T/(m\Omega_E^2)$ for the width of the displacement distribution (4.18) and the equations $\Omega_E = f(\kappa)\Omega_0$ and $\Omega_0 = \sqrt{Q^2/(4\pi\epsilon_0 m\Delta^3)}$ from above follows:

$$\tilde{\Gamma} = \frac{\Delta^2}{\sigma_r^2} = \frac{\Delta^2 m\Omega_E^2}{k_B T} = \frac{\Delta^2 mf(\kappa)^2\Omega_0^2}{k_B T} = \frac{Q^2 f^2(\kappa)}{4\pi\epsilon_0 \Delta k_B T} = f^2(\kappa)\Gamma_{\text{eff}} = \Gamma \quad (5.13)$$

5.5 Interpretation and Discussion

The ratio of areas given by the interparticle distance and the mean displacement of a particle from its equilibrium lattice site, $A_1/A_2 = \Delta^2/\sigma_r^2$, suggested in the beginning of this chapter, offers in fact a simple method for the estimation of the coupling parameter Γ. The essential quantities are easy to measure directly from images of the particle system.

The localized coordinate system introduced for this purpose is very vulnerable to produce artificial features in the case of non-stationary processes due to misinterpretation of the mean lattice sites. In the case of a completely stationary crystal structure, it works well and gives consistent results for Γ in comparison with the particle temperature T, which should both show the same features. The method to apply for the calculation of the displacements is therefore to be chosen according to the present experimental data. Further the spatial resolution can become a restricting factor: the displacements of particles from the mean lattice site (in units of pixels) has to be much larger than the resolvable distance, or the uncertainty in the recorded images. Else the particle either does not seem to move at all, or one could mistake fluctuations in the intensity values due to pixel noise as a particle oscillation.

After these problems are taken into account, the calculation of Δ^2/σ_r^2 is not only a simple way to specify the thermodynamical state of the system, but also the particle charge can be derived, since $\Gamma \propto Q^2/\Delta T$. Necessary for this is the additional estimation of the particle temperature T and the screening parameter κ from the data.

For temperature calculations the temporal resolution must be high: The time step between recorded images has to be small enough to not exceed the time scale of the thermal motion of the particles. On the other hand, the distance (in units of pixels) the particles have to move from frame to frame has to be much larger than the resolvable distance. Here the same reasons apply as mentioned for the displacement above. The particle temperature can then be measured from the width of the velocity distribution.

The factor $F(\kappa)$ introduced in (5.1) describes the contribution of screening to the mean potential energy and is given by the expression $f^2(\kappa)$, which was derived in the last section for several specific structures, in particular for the two-dimensional hexagonal crystal. To derive $f(\kappa)$, κ still has to be estimated, e.g. from the longitudinal branch of the wave spectrum. From (5.13) then follows

$$Q = \frac{\Delta}{\sigma_r} \frac{\sqrt{4\pi\epsilon_0 \Delta k_B T}}{f(\kappa)} \tag{5.14}$$

Charge and Γ are presented in Table 5.3 for experiments I and II. The values for charge Q_{ws} and screening parameter κ_{ws} are those from the wave spectra analysis presented in Sect. 5.1. κ_{ws} was used for the calculation of $f(\kappa)$ and $f_{NN}(\kappa)$ in the nearest neighbor approximation.

From Table 5.3 one can see that Γ and $\tilde{\Gamma}$ are in good agreement, with $\tilde{\Gamma}$ slightly higher than Γ. Questionable stays the use of the nearest neighbor approximation in the calculation of $f(\kappa)$: Γ_{NN} differs by a factor of 1.5–1.6 from Γ and $\tilde{\Gamma}$, respectively. The very similar results for Γ and $\tilde{\Gamma}$ lead to similar particle charges, mutually validating the different approaches of wave spectra and particle motion analysis.

Molecular dynamics (MD) simulations on a two-dimensional system with comparable parameters as the above experiments have been performed and their dynamical properties have been evaluated, presented in [5]. A defect lattice site was included in the simulation. The results were consistent with the experimental findings: the influence of the defect on the local bond order parameter was long-range, while the interparticle spacing and $\tilde{\Gamma}$ were affected only locally by the lattice disruption.

References

1. S. Nunomura, J. Goree, S. Hu, X. Wang, A. Bhattacharjee, Dispersion relations of longitudinal and transverse waves in two-dimensional screened Coulomb crystals. Phys. Rev. E **65**, 0664021–11 (2002)
2. S. Nunomura, J. Goree, S. Hu, X. Wang, A. Bhattacharjee, K. Avinash, Phonon spectrum in a plasma crystal. Phys. Rev. Lett. **89**(3), 035001 (2002)
3. S. Nunomura, S. Zhdanov, D. Samsonov, G.E. Morfill, Wave spectra in solid and liquid complex (dusty) plasmas. Phys. Rev. Lett. **94**, 045001 (2005)
4. U. Konopka, Wechselwirkungen geladener Staubteilchen in Hochfrequenzplasmen. PhD thesis, Fakultät für Physik und Astronomie der Ruhr-Universität-Bochum (2000)
5. C.A. Knapek, A.V. Ivlev, B.A. Klumov, G.E. Morfill, D. Samsonov, Kinetic characterization of strongly coupled systems. Phys. Rev. Lett. **98**, 015001 (2007)
6. B. Liu, J. Goree, V. Nosenko, L. Boufendi, Radiation pressure and gas drag forces on a melamine-formaldehyde microsphere in a dusty plasma. Phys. Plasma **10**(1), 9–20 (2002)
7. A.G. Zagorodny, P.P.J.M. Schram, S.A. Trigger, Stationary velocity and charge distributions of grains in dusty plasmas. Phys. Rev. Lett. **84**(16), 3594–3597 (2000)
8. C. Kittel, Introduction to solid state physics. (Wiley, Toronto, 1976)

Chapter 6
Theory of Phase Transitions in 2D Systems

The phase of a two-dimensional complex plasma can be defined, in the thermodynamical sense, by the Coulomb coupling parameter Γ, the ratio of the average kinetic and potential energy, as it was introduced in Sect. 2.3. Critical values Γ_c at the point of the phase transition between the liquid and solid state in a two-dimensional system have been obtained in computer simulations of the liquid to solid transition of a 2D electron gas as $\Gamma_c = 95 \pm 2$ [1]. Thouless [2] calculated it based on the dislocation mediated melting mechanism as $\Gamma_c = 78$. This melting mechanism will be subject of the following chapter. A later experiment with 2D electron sheets [3] yielded a $\Gamma_c = 137 \pm 15$ which is larger than the theoretical prediction, and close to the critical value for 3D systems of $\Gamma_{c,3D} = 172$ as it was obtained in Monte-Carlo simulations of a one-component plasma (OCP) [4]. The last Chap. 5 gave a simple method for the estimation of Γ of a two-dimensional system of particles, if the particle coordinates are available with a high spatial resolution. Though observing Γ during an experiment involving a phase transition could help to identify the respective phases and the point of phase transition, it gives no information on how this transition might work on a fundamental level. The next chapters will address the different theories attempting to explain the underlying mechanisms.

Several theories for phase transitions of two-dimensional systems exist. They include models of first or second order phase transitions driven by different mechanisms. The next chapters give an overview over the most common theories. Generally accepted is the KTHNY theory [5–9]—named after Kosterlitz, Thouless, Halperin, Nelson and Young. It describes the melting of two-dimensional systems with continuous phase transitions (second order) (Sect. 6.1). In opposition, a first order transition similar to three-dimensional systems is proposed by the grain-boundary theory [10–12] and the density-wave theory [13–15]. The models for first order transitions are explained in Sect. 6.2.

The above models have in common the description of the mechanism of the phase transition on the level of the individual particles, which then can be connected to the thermodynamical quantity, namely the particle temperature defined as the ensemble average over all particles. A modified Lindemann criterion of melting [16] will be

C. A. Knapek, *Phase Transitions in Two-Dimensional Complex Plasmas*,
Springer Theses, DOI: 10.1007/978-3-642-19671-3_6,
© Springer-Verlag Berlin Heidelberg 2011

introduced in Sect. 6.3. It predicts the point of the phase transition using a comparably simple criterion based on the oscillation of particles about their equilibrium positions. The last Sect. 6.4 is concerned with another approach to the mechanism of melting and recrystallization on the kinetic level as it was brought forward by Frenkel [17].

6.1 Second Order Phase Transition: KTHNY

6.1.1 Topological Long Range Order

Kosterlitz and Thouless [5, 6] developed a theory in which the behavior of defects determines the nature of the phase transition, i.e. a theory of dislocation-mediated melting. The consequences of this model for the system properties were examined later by Halperin, Nelson and Young [7–9].

The main achievement of the KTHNY theory was a new definition of long range order of a 2D system based on overall system properties, i.e. the concentration and behavior of dislocations in the system. This is an advantage in cases where the correlation functions, for example a spin-spin correlation function in a 2D spin system, vanish at a finite temperature and can not be used as evidence for the existence of long range order.

In a solid in equilibrium free dislocations do not exist, and the system has a topological long range order. In the liquid state free dislocations which can move along the surface, destroy the long range order. In contrast to the free dislocations, pairs of dislocations with opposite Burgers vector can form due to thermal excitation even at low temperatures in the solid state. They have a finite energy while the energy E_{sd} of a single dislocation in a 2D system increases logarithmically with the system size:

$$E_{sd} = \left(\frac{\nu \Delta^2(1+\tau)}{4\pi}\right) \ln \frac{A}{A_0} \tag{6.1}$$

Δ is the mean interparticle distance, ν and τ are the 2D rigidity modulus and the Poisson ratio, A and A_0 are the system area and the area of a unit cell, respectively. From (6.1) the entropy $S = k_B \ln \frac{A}{\Delta^2}$ and the free energy can be derived. The critical temperature T_{c1} at which its is likely that dislocation pairs split a free dislocation appears can then be calculated as the temperature at which the free energy changes sign:

$$k_B T_{c1} = \nu \Delta^2(1+\tau)/4\pi \tag{6.2}$$

The existence of free dislocations in a system can be determined by tracing a closed path of total length L from one lattice site to the next. To each free dislocations enclosed by the path a Burgers vector can be assigned. The Burgers vector is the vector which is necessary to close an equivalent path with the same number of steps in each direction in an ideal lattice. In the case of more than one dislocations the Burgers vectors are summed up to close the gap. The number of free dislocations N_{dis} is proportional to the enclosed area A. For dislocation pairs N_{dis} is proportional to the path length. Pairs of dislocations contribute only if the path cuts through the pair. The sum of the Burgers vectors is proportional to $\sqrt{N_{dis}}$ (see (6.3)) [5]. This criterion can be to determine the occurrence of topological long range order.

$$\begin{matrix} \text{free dislocations} \\ \text{pairs} \end{matrix} \quad N_{dis} \propto \begin{cases} A \\ L \end{cases}, \quad \bar{B} \propto \begin{cases} \sqrt{A} \\ \sqrt{L} \end{cases} \tag{6.3}$$

Model systems were investigated with regard to the theory including the two-dimensional crystal which is of interest in the case of the two-dimensional plasma crystal. The interparticle correlation function shows no long-range order in that kind of system. Here the displacement of particles from their mean lattice sites is used for energy calculation. The response of the medium with dislocations to stress shows that the energy of isolated dislocations grows logarithmically with the area of the system while the energy stays finite for pairs of dislocations. Therefore dislocations are closely bound for low temperatures.

6.1.2 Consequences of the Kosterlitz–Thouless Approach

Halperin and Nelson applied the Kosterlitz–Thouless approach to a two-dimensional triangular lattice and found a second transition temperature T_{c2} above the formerly mentioned T_{c1}. This makes the melting transition a two-stage process from the solid phase over an intermediate phase—the so called hexatic phase—to the liquid. The consequences for structural properties of the system are investigated and predictions for the long range behavior of the pair and bond correlation functions are made.

The pair correlation function $g(r)$ is a measure of the translational order of a system. It gives the probability to find a particle at a distance r from another one:

$$g(r) = \sum_{r-\delta r \leq r < r+\delta r} \langle \rho(\mathbf{r})\rho^*(\mathbf{0}) \rangle \tag{6.4}$$
$$\rho(\mathbf{r}) = \exp\{i\mathbf{G}[\mathbf{r} + \mathbf{u}(\mathbf{r})]\}$$

where $\mathbf{G}$ is the reciprocal lattice vector and $\mathbf{u}$ the displacement of a particle form its mean lattice site.

Similar, the bond correlation function $g_6(r)$ for the orientational order can be defined. It measures the orientation of nearest neighbor bonds separated by the distance r.

$$g_6(r) = \sum_{r-\delta r \leq r < r+\delta r} \langle \Psi^*(\mathbf{r})\Psi(\mathbf{0}) \rangle \tag{6.5}$$
$$\Psi = \exp\{i\theta(\mathbf{r})\}$$

with the angle $\theta(\mathbf{r})$ between a nearest neighbor bond and an arbitrary chosen axis.

In the solid phase $T < T_{c1}$ all dislocations are bound in pairs. The pair correlation function $g(r)$ decays with a power law $\propto r^{-\eta(T)}$ for large distances r due to fluctuations of the particle displacements [18]. There is no long range translational order in the system. The bond correlation function $g_6(r)$ tends to a nonzero constant for large r, therefore there is long range orientational order [19]. This definition of a solid differs from that of a three-dimensional solid due to the lack of translational order. Still the state is clearly different from a liquid.

As the temperature reaches T_{c1} from below, an upper bound for $\eta(T)$ is found which is $1/3$ for a triangular lattice. At T_{c1} the dislocation pairs start to dissociate, and for $T > T_{c1}$ $g(r)$ decays exponentially $\propto e^{-r/\xi}$ with a correlation length ξ. Orientational order persists with a slow algebraic decay of $g_6(r) \propto r^{-\eta_6(T)}$. This transition if often referred to as the Kosterlitz–Thouless transition.

Halperin and Nelson found a second transition at a temperature $T_{c2} > T_{c1}$. Here the dislocations themselves break up into free disclinations and both correlation functions decay exponentially for $T > T_{c2}$ [7, 8]. The following shows a compilation with the most important results:

$$
\begin{array}{lll}
\text{solid} & T < T_{c1} & \left\{ \begin{array}{l} g(r) \propto r^{-\eta(T)} \\ g_6(r) \to \text{const. for large } r \end{array} \right. \\
 & T \to T_{c1}^- & \eta(T) \to \text{upper bound (1/3 for triangular lattice)} \\
\text{hexatic} & T_{c1} < T < T_{c2} & \left\{ \begin{array}{l} g(r) \propto e^{-r/\xi} \\ g_6(r) \propto r^{-\eta_6(T)},\ \eta_6(T) = 18k_B T/(\pi K_A(T)) \end{array} \right. \\
 & T \to T_{c1}^+ & \xi_+(T) \propto \Delta \exp\left\{ \dfrac{b}{(T/T_{c1}-1)^{0.36963}} \right\} \\
 & T = T_{c2} & \eta_6(T_{c2}) = 0.25 \\
\text{liquid} & T_{c2} < T & \left\{ \begin{array}{l} g(r) \propto e^{-r/\xi} \\ g_6(r) \propto e^{-r/\xi_6},\ \xi_6(T) \propto \exp\left(a/|T-T_{c2}|^{0.5}\right) \end{array} \right.
\end{array}
$$

with b, a constants, Δ is the lattice constant and $K_A(T)$ is the Franck constant which is $\propto \xi_+^2(T)k_B T$. It goes to infinity shortly above T_{c1}, and jumps to zero at T_{c2}. The exponent 0.36963 in the proportionality of ξ has been derived by Young [9] under consideration of angular and radial interactions between dislocation pairs in a triangular lattice.

6.1.3 Summary of KTHNY

The KTHNY theory describes a defect mediated melting with two transition points for two-dimensional systems. Starting with a solid phase, the first transition involves the breaking up of bound dislocation pairs. An intermediate phase appears with no long range translational order, but still orientational order can be found in the system. This phase is referred to as the hexatic phase. At the second transition point at a higher temperature the dislocations dissociate into free disclinations and the liquid state is reached. Both transitions are continuous (2nd order) phase transitions. The phases can be characterized by topological properties solely. Further, the theory makes specific predictions about the behavior of the translational and orientational correlation functions which can be used to compare the theory with experimental data.

6.2 First Order Phase Transition

Two other approaches involving a first order transition for two-dimensional systems are worth mentioning. The density wave theory by Ramakrishnan and Yussouff [13] describes the freezing of a two-dimensional system as a first order transition without an intermediate phase. Melting due to the generation of grain boundaries is addressed in the second theory by Chui [12].

6.2.1 Density-Wave Theory

The density-wave theory [13, 14] is a mean field theory for two-dimensional systems. If a density wave forms in the system near the melting temperature, the positional correlation is lost. From the density change the free energy balance between solid and liquid is derived. This leads to a freezing condition which is independent of interaction forces between the particles (and therefore of the crystal structure). Only short-range two-body correlations and geometrical factors enter the calculation [13].

Monte Carlo simulations at a fixed pressure at different temperature [15] showed a discontinuity in the density as well as the occurrence of latent heat. This points to a first order transition between the solid and liquid phase in contrast to the predictions of the KTHNY theory. Further no hexatic phase could be found in the simulations.

As Ramakrishnan points out, there is still the possibility that the first order transition takes place before or after the dislocation transition, but with long relaxation times so that the system seems to be solid within the observation time.

One problem of the density-wave theory was addressed in [20]: As a mean-field theory it neglects the fluctuations that destroy the long range positional order and might not be applicable to predict the nature of a transition.

6.2.2 Grain-Boundary Theory

Chui [11, 12] proposes a melting transition that appears due to the spontaneous generation of grain boundaries at a lower temperature than the one where dislocations unbind (T_{c1} in the above KTHNY description). The grain boundaries consist of chains of dislocations, therefore the number of dislocations has to show a sharp increase at the point of melting. A melting scenario due to grain boundaries was already considered earlier in [10] for small-angle grain boundaries, but it was concluded there to be identical to the KTHNY mechanism.

In opposition, Chui's calculations lead to a first order transition, if the grain boundaries are coupled to a finite change in density or to bound dislocation pairs. The quantity of the core energy of the dislocations [10] plays an important role: For core energies below $2.84T_{c1}$, he predicts that the transition goes from weakly first-order to strongly first-order. As in the density-wave theory, the hexatic phase does not exist as long as there are no bound states between grain boundaries with opposite Burgers vector. In the case of bound states, the Burgers vectors of grain boundaries could cancel out, and there would be orientational long range order, defined by a power-law decay of $g_6(r)$. Without bound states, $g_6(r)$ decays exponentially and the hexatic phase can exist, but only for temperatures smaller than T_{c1} [12].

According to [20], for dislocation core energies $< 2.8T_{c1}$ the distance between dislocations constituting to a grain boundary become large compared to the distance between grain boundaries (which opposes the assumptions of Chui), and the predictions of a first-order transition for core energies below $2.8T_{c1}$ are not reliable.

6.3 Lindemann Criterion of Melting

Lindemann stated [16] that the melting of a three-dimensional crystal appears when thermal vibrations of the particles around their mean lattice sites get large enough for neighboring particles to collide. Later [21] this criterion was stated more precisely: When the root mean amplitude $\sqrt{\langle u^2 \rangle}$ of the vibrations reaches $\approx 10\%$ of the mean interparticle distance, the solid melts. This criterion is not applicable to two-dimensional systems, because the fluctuations $\sqrt{\langle u^2 \rangle}/\Delta$ become infinite as the system size goes to infinity.

An approach to estimate a critical temperature in two-dimensional systems based on Lindemanns work was done by Lozovik and Farztdinov [22]. A dimensionless parameter γ is introduced as the ratio of the mean square difference of particle displacements $\mathbf{u}$ from their mean lattice sites to the squared interparticle distance:

$$\gamma = \frac{\langle (\mathbf{u}_i - \mathbf{u}_{i+1})^2 \rangle}{\Delta^2} \tag{6.6}$$

This leads to a Lindemann-like criterion of melting as γ reaches a critical value γ_c which was confirmed to be valid by molecular dynamics simulations by Bedanov et al. [23]. They find a linear γ below the melting temperature, and a sharp growth at the melting point which coincides with leaps in the parameters describing translational order, while orientational order persists until a higher temperature. This is consistent with the results of the KTHNY theory. For Coulomb systems a $\gamma_c \approx 0.1$ was found [23, 24] for the melting of a two-dimensional system from the solid to the hexatic phase.

6.4 Kinetic Theory of Melting (Frenkel)

In his book 'Kinetic Theory of Liquids' [17], Frenkel describes the process of melting on the kinetic level.

The crystalline state is defined by a regular lattice with fixed equilibrium positions for each particle. At any finite temperature, the particles are subject to heat motion, which in turn causes particle vibrations around the equilibrium positions. If the energy of a particle is high enough to overcome the potential barrier set by the neighboring particles, it can jump from its lattice site into an adjacent interstice, leaving a hole in the regular structure. Both holes and interstices can independently diffuse through the crystal.

The formation of such a lattice defect increases the internal energy of the whole system by the amount of the activation energy U_D the particle needed to jump into the interstitial position.

Then the free energy of N_D defects becomes $F = N_D U_D - TS$ where the entropy S is given by the number of possibilities P_D to distribute N_D defects in a system of N particles with a regular lattice structure.

Assuming distinguishable particles, the probability is $P_D = N!/N_D!(N - N_D)!$ and the entropy is $S = k_B \ln P_D$. The minimum of $F(N_D)$—corresponding to the thermodynamical equilibrium—is given by $\partial F/\partial N_D = 0$ and leads to the relation:

$$N_D = N \exp\left(-\frac{U_D}{k_B T}\right) \tag{6.7}$$

The total number of defect sites decreases exponentially with the temperature of the particles (Arrhenius law). A higher activation energy U_D will decrease the slope of the decay.

Recently evidence for this behavior has been found [25] in a two-dimensional complex plasma system observed in different kinetic states. In that experiment, the particles were heated by two incident laser beams. Depending on the laser power, steady state regimes of certain particle temperatures could be obtained. The number of defects identified in the system for each laser power decayed exponentially with the particle temperature.

The transition from the solid to the liquid state is accompanied by a volume increase of 10% at the transition point, which is small enough so that the arrangement of particles in the liquid must be similar to the solid, but large enough to allow particles to be displaced and constitute to a high fluidity. The character of heat motion in the liquid close to the crystallization point is fundamentally the same, consisting of small vibrations around equilibrium positions, but the mean time of a particle remaining on such a position is much shorter as in the solid.

The main difference lies in the degree of order: in the solid state, long range transitional and orientational order exist, while in the liquid the order is lost in the system due to the stronger heat motion and fluidity. The effect of the heat is amplified by the increase of volume which in turn lowers the cohesive forces.

Frenkel also described the behavior of liquid crystals, which, when heated from the solid state, first loose their translational long range order, but keep a high degree of orientational order within larger continuous groups of particles ("swarms"). This intermediate liquid-crystalline state will melt to the liquid state at further increase of the temperature by gradually decreasing the orientational order and replacing the large swarms by smaller, so-called "cybotactic", groups of particles. An expression for the average number of molecules in such a group or domain in dependence of the temperature of the system can be derived by the following considerations.

If the total number of molecules is N, then there are $z = N/\bar{N}_D$ homogenous regions each containing N_D molecules. While the molecules are oriented equal within one region, the orientation of two domains can be completely different. Each region has a surface area, resulting in an increase of the internal energy of the system due to the additional surface energy by $E = \sigma V^{2/3} z^{1/3}$ with the surface tension σ and the enclosed volume V. The distribution of molecules across the domains is described by the probability $P = N!/[(N/z)!]^z \approx z^N$ which increases the entropy of the system by $S = k_B \ln P = k_B N \ln z$. Then in equilibrium it follows from $\partial F/\partial N_D = 0$ that $z = \left(\frac{3Nk_BT}{\sigma V^{2/3}}\right)$ and

$$N_D = \frac{N}{V}\left(\frac{\sigma V}{3Nk_BT}\right)^3 \tag{6.8}$$

In this derivation the behavior of structural order parameters are not discussed specifically, but it is to be expected that, due to the domain forming, long range orientational order can not persist.

The theoretical derivation, originally derived for the behavior of molecule crystallites, fits very well as a model for the behavior of the rapidly cooling two-dimensional complex plasma and is able to explain the experimental findings presented in the following chapters more conveniently than other theories.

References

1. R.W. Hockney, T.R. Brown, A lambda transition in a classical electron film. J. Phys. C **8**, 1813–1822 (1975)
2. D.J. Thouless, Melting of the two-dimensional Wigner lattice. J. Phys. C **11**, L189–L190 (1978)
3. C.C. Grimes, G. Adams, Evidence for a liquid-to-crystal phase transition in a classical, two-dimensional sheet of electrons. Phys. Rev. Lett. **42**(12), 795–798 (1979)
4. D.H.E. Dubin, First-order anharmonic correction to the free energy of a coulomb crystal in periodic boundary conditions. Phys. Rev. A **42**(8), 4972–4982 (1990)
5. J.M. Kosterlitz, D.J. Thouless, Long range order and metastability in two dimensional solids and superfluids. J. Phys. C **5**, 124–126 (1972)
6. J.M. Kosterlitz, D.J. Thouless, Ordering, metastability and phase transitions in two-dimensional systems. J. Phys. C **6**, 1181–1203 (1973)
7. B.I. Halperin, D.R. Nelson, Theory of two-dimensional melting. Phys. Rev. Lett. **41**, 121 (1978)
8. D.R. Nelson, B.I. Halperin, Dislocation-mediated melting in two dimensions. Phys. Rev. B **19**, 2457 (1979)

9. A.P. Young, Melting and the vector Coulomb gas in two dimensions. Phys. Rev. B **19**, 1855 (1979)
10. D.S. Fisher, B.I. Halperin, R. Morf, Defects in the two-dimensional electron solid and implications for melting. Phys. Rev. B **20**(11), 4692–4712 (1979)
11. S.T. Chui, Grain-boundary theory of melting in two dimensions. Phys. Rev. Lett. **48**(14), 933–935 (1982)
12. S.T. Chui, Grain-boundary theory of melting in two dimensions. Phys. Rev. B **28**(1), 178–194 (1983)
13. T.V. Ramakrishnan, M. Yussouff, First-principles order-parameter theory of freezing. Phys. Rev. B **19**(5), 2775–2794 (1979)
14. T.V. Ramakrishnan, Density-wave theory of first-order freezing in two dimensions. Phys. Rev. Lett. **48**(8), 541–545 (1981)
15. F.F. Abraham, Melting in two dimensions is first order: an isothermal-isobaric Monte Carlo study. Phys. Rev. Lett. **44**((7), 463–466 (1980)
16. F.A. Lindemann, Über die Berechnung molekularer Eigenfrequenzen. Physik. Zeits. **11**, 609–612 (1910)
17. J. Frenkel, *Kinetic Theory of Liquids*. (Dover Publications, Inc., New York, 1955)
18. B. Jancovici, Infinite susceptibility without long-range order: the two-dimensional harmonic "solid". Phys. Rev. Lett. **19**(1), 20–22 (1967)
19. N.D. Mermin, Crystalline order in two dimensions. Phys. Rev. **176**(1), 250–254 (1968)
20. K.J. Strandburg, Two-dimensional melting. Rev. Mod. Phys. **60**(1), 161–207 (1988)
21. J.J. Gilvarry, The Lindemann and Grüneisen laws. Phys. Rev. **102**(2), 308–316 (1956)
22. Yu.E. Lozovik, V.M. Farztdinov, Oscillation spectra and phase diagram of two-dimensional electron crystal: "new" (3+4)-self-consitent approximation. Solid State Commun. **54**(8), 725–728 (1985)
23. V.M. Bedanov, G.V. Gadiyak, On a modified Lindemann-like criterion for 2D melting. Phys. Lett. **109**((6)), 289–291 (1985)
24. X.H. Zheng, J.C. Earnshaw, On the Lindemann criterion in 2D. Europhys. Lett. **41**(6), 635–640 (1998)
25. V. Nosenko, S.K. Zhdanov, A.V. Ivlev, C.A. Knapek, G.E. Morfill, 2D melting of plasma crystals: equilibrium and nonequilibrium regimes. Phys. Rev. Lett. **103**(1), 015001 (2009)

Chapter 7
Recrystallization of a 2D Plasma Crystal

A phase transition in a complex plasma can be induced by several mechanisms, e.g. laser induced heating of the crystal [1–4], changing of plasma parameters (pressure change, rf power changes) [5–7], or electric manipulation [8, 9]. Either a crystalline system can be melted and thus brought into a liquid or gaseous state, or the other direction, i.e. the recrystallization of a unordered system can be investigated. In any case, the process has to be observed at a high temporal resolution to obtain dynamical properties, and a good spatial resolution to derive the structural properties mentioned in the last chapter in theoretical models, and to compare them.

Several experiments were performed with the aim to observe the process of recrystallization and to characterize it by connecting the thermodynamical state with the particle properties on a kinetic level. Part of this work has been published in [9].

The recrystallization was induced by a short electric pulse which melted a two-dimensional complex plasma, initially in a highly ordered state. Images of the initial state, the melting and the following recrystallization were recorded with a high speed digital camera. The experimental parameters of two experiments which will be presented in detail are given in Sect. 7.1. That section also covers the technical details of the induced melting.

The thermodynamical state of the system is determined by the particle temperature, which can be obtained from the analysis of the particle oscillations at their mean lattice sites, as it was already done in Chap. 5. To resolve this motion, a high spatial and temporal resolution is required. Since the temperature is a quantity of the particle ensemble rather than of individual particles, a large number of particles have to be available for valid statistical averaging. In an experiment one has to find a compromise between those two requirements. In the presented experiments the spatial resolution was chosen rather low to obtain a large number of particles in the field of view. The measurement uncertainties and the implications on the particle temperature analysis will be discussed in Sects. 7.2 and 7.3 in detail.

The global and local structural properties of the system are investigated in Sects. 7.5 and 7.6. Finally a connection between the thermodynamical state of the system and the structural changes it undergoes during recrystallization is established

C. A. Knapek, *Phase Transitions in Two-Dimensional Complex Plasmas*,
Springer Theses, DOI: 10.1007/978-3-642-19671-3_7,
© Springer-Verlag Berlin Heidelberg 2011

in Sects. 7.7 and 7.8.1. The last Sects. 7.8.2 to 7.9 contain a theoretical approach to explain the experimental findings, and the interpretation of the results in the context of the theories.

7.1 Experimental Parameters

Two experiments will be presented in the following, referred to as experiments rI and rII. The basic setup for the experiments was described in Sect. 3.1. The parameters used here are very similar to those used in the experiments to measure Γ in Chap. 5 The main differences lie in the larger number of particles in the field of view at the cost of a lower spatial resolution, and by the manipulation of the particles by an electric pulse. All experimental parameters and settings are listed in Table 7.1.

An argon plasma was ignited with the radio-frequency power set to 10 W forward and 0 W reflected, with a peak-to-peak voltage measured between driven electrode and ground of -172 V. The self bias at the lower electrode was measured to be -70 V and the neutral gas pressure was 1.938 Pa at a flow rate of 4 sccm. Two parallel tungsten wires were mounted horizontally inside the chamber at a height of 8.2 mm above the lower electrode. The gap between the wires was 58.7 mm (see Fig. 3.3).

Melamine-formaldehyde (MeF) particles with a diameter of $9.19 \pm 0.09\,\mu$m were inserted by shaking the particle dispenser. The particles arranged themselves in a horizontal layer 8 mm above the electrode, approximately 0.2 mm below the level of the wires. The crystal extended over an estimated area of 69.6×38.5 mm^2 between the two wires. The system was checked for its two-dimensionality by moving the illuminating laser and the camera vertically over the extend of the particle plane and it was verified that above or below the layer no particles were to be seen in the live stream of the camera.

The illuminating laser was set to a power of 132 mW. The CMOS high speed camera with a 105 mm lens achieved a spatial resolution of 0.034 mm/px with an image size of 1024×1024 px^2 corresponding to a region of 34.8×34.8 mm^2. Each set of recorded images contains 6144 images at a frame rate of 500 fps, yielding a total measurement time of 12.29 s per run.

The function generator attached to the wires was set to generate a pulse every 100 s, with a carrier frequency of 1 Hz and a 20% duty cycle. The pulse length is then (duty cycle)/(carrier frequency) = 0.2 s . For the duration of the pulse, a 5 V peak-to-peak voltage was applied to a circuit which then opened the connection between the power supply, set to -253 V, and the wires.

The function generator also triggered the camera to start a record at an adjustable time before the next pulse, usually 2 s. To avoid repeated melting of the crystal and thus destabilize it, the power supply was turned down except for the time when a recording was done.

A schematic view through the top window of the chamber is shown in Fig. 7.1. The wires are drawn on the top and bottom with the two-dimensional particle cloud

Table 7.1 Experimental parameters for the recrystallization experiments rI and rII

Discharge settings	
rf power forward/reflected	10/0 W
Self bias	-70 V
Peak-to-peak voltage	-172 V
Gas	Ar
Flow rate	4 sccm
Neutral gas pressure	1.938 Pa
Particles	
Particles	MeF, $\varnothing\, 9.19 \pm 0.09\,\mu$m
Particle mass density	1.51 g/cm^3
Particle mass	6.14×10^{-13} kg
Average interparticle distance	0.6 mm
Epstein drag coefficient	2.28 Hz ($\delta = 1.26$), 2.68 Hz ($\delta = 1.48$)
Particle charge	-12200 ± 340 e
Screening parameter	0.77 ± 0.03
Height of particles above the electrode	8.0 mm
Recording	
Recorded frames	6144
Frame rate	500 fps
Start of recording (before pulse)	2 s
Resolution	0.034 mm/px
Illumination laser power	132 mW
Pulse settings	
Height of wires above the electrode	8.2 mm
Voltage at the wires during pulse	-253 V
Pulse duration	0.2 s

The derivation of the particle charge, screening parameter, the Epstein drag coefficient and the average interparticle distance will be done in Sects. 7.3 and 7.5.1. The two values for the Epstein coefficient are calculated for the two different reflection indices $\delta = 1.26$ and $\delta = 1.48$ from the references [10] and [11]

(shaded area) located in between. The dashed rectangle indicates the field of view of the camera. The coordinate system is drawn to the right for later reference of the x, y coordinates used in the analysis. The dotted horizontal line marks the center between both wires and divides the field of view in two regions A, B. The arrows indicate the particle motion when the pulse is applied: both wires accelerate the negatively charged particles abruptly towards the center region, and the crystalline order is completely destroyed. When reaching the central region, particles from both sides collide and reverse their motion. Due to the strong damping, the system relaxes to a crystalline state very fast as soon as the pulse ends.

The induced common motion of particles is always opposite in the regions marked A and B in Fig. 7.1. To subtract this directional motion perpendicular to the center line, particles in the region A and B are analyzed separately in the initial statistical

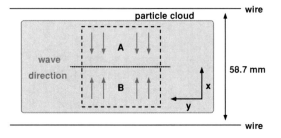

Fig. 7.1 Schematic view from the top. The wires used for excitation were mounted parallel at nearly the same height as the particle layer. The *arrows* indicate the direction into which the particles are pushed when a negative pulse is applied to both wires simultaneously. They meet in the center at the *dotted line*, dividing the field of view into the regions A and B. The *dashed rectangle* marks the field of view of the camera. The coordinate system shows the notation used for the images axis later in the analysis

analysis, i.e. the velocity distributions. Quantities like the velocity dispersion which are not affected by the common motion can then be averaged over both regions.

During and shortly after the pulse the particles move too fast to be traced from frame to frame. Further, particles might leave the field of view for a short time after the pulse impact as the system relaxes towards the wires and particles shoot past their initial positions.

The data presented in later chapters are derived for times after the melting when the number of particles in the field of view was at least 80% of the average undisturbed number of particles, which was measured in the 2 s before the pulse. Also, particles had to be traced for at least 30 contiguous frames to be taken into account. Figure 7.2 shows the number N_P of tracked particles vs. time meeting this criterium. In both experiments the pulse hits at 2 s in the recording time, characterized by the sharp drop of N_P. The fluctuations at $t \approx 4$ s is caused by the rebound of the particles towards the center. Note that this is not the actual number of particles seen by the eye in the images, but only the number of particles which could be traced for at least 30 frames. In the later analysis, the time between 2 and 3.5 s was omitted.

In experiment rI, the average number of particles used for the data analysis was close to 1900, while in experiment rII a larger region of the images could be used, including approximately 2700 particles. The region of the images suitable for analysis is in general restricted by the weaker illumination along the two edges parallel to the wires (i.e. along the y-direction close to both wires in Fig. 7.1). The reason seems to be a variation in the particle levitation heights from the edges parallel to the wires towards the center (in x-direction), making it difficult to obtain an uniform illumination. This could be caused by a slightly curved potential due to the permanent weak negative floating potential of the wires. Depending on the case, the original field of view supplied by the camera was cut in the analysis to omit badly illuminated particles from the analysis. In experiment rI the effective area became $623 \times 823 \, \text{px}^2$, while in experiment rII is was $773 \times 1023 \, \text{px}^2$.

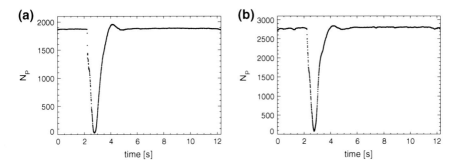

Fig. 7.2 Number N_P of particles per frame. Only particles which could be traced from frame to frame in at least 30 contiguous frames were counted. The melting pulse hits at 2 s, from that on the analysis was discontinued until 3.5 s. **a** Experiment rI, **b** Experiment rII

7.2 Uncertainties

Particle coordinates and velocities were extracted from the images. An analysis of the expected uncertainties due to tracking inaccuracies and pixel noise was performed by comparing the number of illuminated pixels per particle and the quantity of the pixel-noise with the results of the error analysis given in Chap. 9. The pixel-noise level is estimated from intensity fluctuations of the particle images and is approximately 7 in units of intensity (the maximum intensity is 255). The average size of the particles lies in the range 6–7 pixels per particle, obtained as the mean of a Gaussian fitted to the distribution of the number of pixels per particle for each experiment. The specific values are listed in Table 7.2 along with the deduced uncertainty in the absolute particle positions. The uncertainties are also given for a reduced framerate of 500/3 fps obtained by averaging each three consecutive images. In that case, the pixel-noise level is reduced to $1/\sqrt{3}$ of the original noise due to its Gaussian nature: averaging leads to a convolution of the Gaussian noise of three images, resulting in a Gaussian noise with width $\sqrt{3}$ times the original noise width, but divided by three due to averaging.

Note that reducing the frame rate has practically no influence on the quantity of the error of the absolute coordinates. It gains importance for the magnitude of the uncertainty in quantities depending on the displacements from frame to frame, e.g. the velocities and from that the particle temperature. It was already mentioned in earlier chapters that the error of displacements depends on the value of the displacement itself. Figure 7.3 illustrates this dependency for the particle sizes and noise levels of interest here. Figure 7.3a shows the situation for a frame rate of 500 fps, Fig. 7.3b for the reduced 500/3 fps.

The intersection of the error curve with the black bisecting line at ≈ 0.05 px (500 fps) and 0.03 px (500/3 fps) gives the lower limit to any measureable distance (blue dashed line). Translated to temperatures measured from velocity distributions, the full framerate of 500 fps is restricted to $T > 2.5$ eV while for $500/3 \approx 166$ fps the

Table 7.2 Pixels per particle, pixel-noise level and errors of the absolute coordinates for experiments rI and rII

	rI	rII
	500 frames per second	
Number of pixels/particle	6.655 ± 0.001	7.853 ± 0.001
Pixel-noise level	6.9	7.1
Error absolute coordinates $\delta_{r,abs}$ [px]		
	0.073	0.075
	500/3 frames per second	
Pixel-noise level	4.0	4.1
Error abs. coord.$\delta_{r,abs}$ [px]	0.070	0.063

Values are given for the original frame rate of 500 fps, and for a reduced frame rate of $500/3 \approx 166$ fps, obtained by averaging each three consecutive images

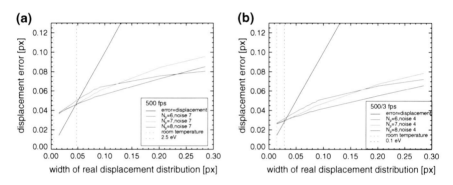

Fig. 7.3 Displacement error vs. real particle displacement as measured for artificially generated particles. The artificial particles were created as cummulations of pixels with a Gaussian intensity profile. They were given a temperature by adding a random velocity drawn from a Gaussian distribution to the former defined particle positions sucessive for several time steps. Pixel noise was added to these artificial images, and the deviation of the tracking and tracing results from the real, known values was calculated. The *black solid line* is the bisecting line (error = displacement), the *colored curves* are the errors for special cases of pixels/particle and noise-levels (see legend). The *vertical dashed red line* marks the expected displacment of particles at room temperature, the *dashed blue line* is the intersection from which on the error becomes smaller than the real displacement. **a** Pixel-noise 7 for 500 fps; **b** Pixel-noise 4 for 500/3 fps

limit is at 0.1 eV. Room temperature 0.025 eV could not be measured until lowering the frame rate down to ≈70 fps.

The reduced frame rate of 166 fps is preverable, since it still provides a high temporal resolution, but lowers the errors to a tolerable value, therefore from now on, the notation one frame will refer to the overlay of three original frames. The next chapter will jusitfy this procedure with regard to some physical aspects.

7.3 Time Scales

The finite particle temperature causes the particles to oscillate with the Einstein frequency Ω_E about their equilibrium lattice sites located within their nearest neighbor cage. According to the Nyquist-Shannon sampling theorem, the sampling rate (in this case the frame rate of the camera) must be larger than two times the lowest frequency to be measured:

$$\text{frame rate} > 2\Omega_E$$

To estimate the Einstein frequency, the relation between the dispersion of the displacements from the mean lattice site, σ_r, and Ω_E could be used: $\Omega_E = \sqrt{k_B T/(m\sigma_r^2)}$. But during recrystallization the particle system in the experiments rI and rII is dominated by non-stationary processes, which makes it difficult to determine a mean lattice site and σ_r. Further it is questionable if the real particle temperature $k_B T$ can be estimated due to the restrictions imposed by the low spatial resolution and the resulting uncertainties, as was shown in the former Sect. 7.2.

If the particle charge and the screening parameter κ are known, Ω_E can be estimated using (5.11) as $\Omega_E = \Omega_0 f(\kappa)$ with $f(\kappa) = (a + b\kappa^2)^{-1}$ calculated from the polynomial fit to the curve in Fig. 5.12.

The last 546 frames of experiments rI and rII, and of two additional experiments where used for a wave spectra analysis to obtain the particle charge Q and κ. All experiments where performed at the same discharge and pressure settings. First, for each set of images the spectrum analysis was performed separately. The individual spectra matrices were then added up to achieve a combined spectrum with a higher resolution. The fitting of the transversal and longitudinal sound velocities as described in Sect. 4.1 gives a particle charge $Q = 12200 \pm 340\,e$ and $\kappa = 0.77 \pm 0.03$. With the average interparticle distance $\Delta = 0.6\,\text{mm}$ derived later (Sect. 7.5.1), this yields:

$$f(\kappa) = 2.25 \pm 0.01$$
$$\Omega_0 = 16.1 \pm 0.4\,\text{s}^{-1} \qquad \Rightarrow \Omega_E = 36 \pm 1.0\,\text{s}^{-1}$$

The minimum sampling rate allowed is therefore $2 \times 36/(2\pi) \approx 11\,\text{Hz}$, meaning a maximum time step of $dt = 0.09\,\text{s}$.

Depending on the planned analysis, dt is further restricted: Assuming room temperature of $0.025\,\text{eV}$ and purely thermal motion, the width of the velocity distribution would be $\sigma_v = \sqrt{k_B T/m} = 0.081\,\text{mm/s}$ with the particle mass $6.14 \times 10^{-13}\,\text{kg}$. The width of the displacement distribution would then be $\sigma_r = \sigma_v/\Omega_E = 0.0028\,\text{mm}$. Again assuming a particle oscillates from $-\sigma_r$ to $+\sigma_r$ with respect to the mean lattice site, and one wanted to reduce the frame rate as much as to just measure that distance $\langle \Delta x \rangle = 0.0028\,\text{mm}$, it applies for $dt = \langle \Delta x \rangle/\sqrt{k_B T/m} = 0.033\,\text{s}$ matching a frame rate of $\approx 30\,\text{Hz}$. At frame rates smaller than that, the particle will already be on its way back in the oscillation path. To measure more than just one point of each period, the frame rate should be considerably higher that 30 fps. Note that according to the last chapter, only temperatures $T > 0.1$ eV can be measured.

Another time scale of interest for the particle dynamics is defined by the Epstein damping coefficient—the damping rate of particle motion due to the collisions of the particle with neutral gas atoms. It was calculated from known gas parameters and the particle size and mass [12, 10] using (4.13). For Argon gas at room temperature at a pressure of 1.94 Pa and particles with a radius of 4.595 μm and a mass of 6.14×10^{-13} kg, the damping rate is $\nu_{Ep} = 2.28$ Hz for the reflection index $\delta = 1.26$ measured in [10], or $\nu_{Ep} = 2.68$ Hz for $\delta = 1.48$ from [11]. This defines a time scale of 0.44 s or 0.37 s, respectively.

The requirement on the time step dt between two frames is (with the smaller value from the damping rate)

$$dt < \left(\frac{\langle \Delta x \rangle}{\sqrt{k_B T/m}} = 0.033\,\text{s} \right) < \left(\frac{\pi}{\Omega_E} = 0.09\,\text{s} \right) < \left(\frac{1}{\nu_{Ep}} = 0.37\,\text{s} \right) \quad (7.1)$$

With the frame rate of 166 fps chosen in the last chapter to reduce the measurement uncertainties, $dt = 0.006$ and the above condition is met more than sufficiently.

7.4 Particle Kinetic Energy

A consequence of the measurement uncertainties described in Sect. 7.2 is the impossibility to resolve real particle temperatures below a certain level in the presented experimental data. It is safe to assume that the measured velocities represent the kinetic energy of the particles which might be composed not only of the heat motion, but also of motion caused by other forces. To which extent the kinetic energy differs from the real temperature is unknown, but since collective particle motions will be subtracted from the velocities, the behavior of the mean kinetic energy should reflect the thermodynamical state of the system well enough.

The histograms of particle velocities used in the following were Gaussian distributed, as it was found in other experiments (e.g. [13, 1]), and the bin sizes were chosen to be larger than the measurement uncertainty to even out statistical fluctuations. For each frame the mean particle kinetic energies E_x, E_y were obtained by using the standard deviations of Maxwellian fits to the velocity histograms of v_x and v_y. By analyzing the regions A and B, labeled in Fig. 7.1, separately, for each frame a set of four distribution functions $p(v_x)_A$, $p(v_y)_A$, $p(v_x)_B$ and $p(v_y)_B$ is obtained. Examples of those functions are shown in Fig. 7.4 together with the corresponding fit. The position of the mean of the distributions differs especially for the x-directions in regions A and B due to the opposite sign of the collective particle motion. Without splitting the field of view, the distributions would become non-Maxwellian with two peaks and lead to wrong fitting results.

The quality of the Maxwellian fit is given by its reduced χ^2, which was on average 1.3 for experiment rI and 1.5–2.2 for experiment rII. The kinetic energies in x- and y-direction are given by the average of the value for both regions A and B:

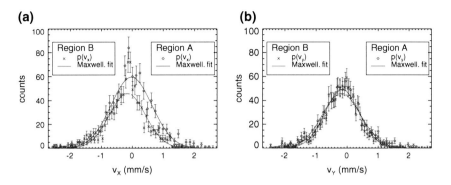

Fig. 7.4 Examples of velocity distributions for the regions A (*blue diamonds*) and B (*red crosses*) indicated in Fig. 7.1. The *error bars* are derived from Poisson weighting as $\sqrt{p(v)}$. *Solid lines* are Maxwellian fits. **a** $p(v_x)$; **b** $p(v_y)$

$$k_B E_x = 0.5(\sigma^2_{vx,A} + \sigma^2_{vx,B})m \qquad (7.2)$$

$$k_B E_y = 0.5(\sigma^2_{vy,A} + \sigma^2_{vy,B})m \qquad (7.3)$$

E_x and E_y are plotted in Fig. 7.5a, b for the experiments rI and rII. They show the same characteristics during the recrystallization process. The kinetic energy returns to the constant value of ≈ 0.1 eV also found in the initial state at 0–2 s. The limit to the measurable energy given by the error is indicated by the solid black line in Fig. 7.5. It is clear, that the constant energies of 0.1 eV are of the magnitude of the measurement error between 0 and 2 s and from approximately 8 s (rI) respectively 10 s (rII) on. Here it is not possible to separate the real σ_v from the noise distribution. However, the data of interest lie in the region in between, where the recrystallization takes place. The errors δE_x, δE_y are taken from the 1-σ uncertainty of the fit parameter distribution width calculated by the fit procedure. The magnitude of those errors depends on the goodness of the fit. To avoid an overloaded plot, only one representative error bar is shown in Fig. 7.5. The relative errors $\delta E_{x,y}/E_{x,y}$ were constant at $\approx 3.8\%$ in rI and $\approx 3.3\%$ in rII.

The energies in x- and y-direction are averaged for each frame at a time t to obtain the kinetic energy $E(t) = [E_x(t) + E_y(t)]/2$. The behaviour of $E(t)$ represents the exponential decay of its components, and fits $E(t) = ce^{-vt} + \langle E \rangle$ were performed with a constant c, the slope v and a ground value $\langle E \rangle$ which, in the ideal case, would be the ground energy of the particle system at room temperature. Here, it will reflect the limit imposed by the uncertainties.

Experiment rI Figure 7.6a shows $E(t) - \langle E \rangle$, $\langle E \rangle = 0.1$ eV, for the part of the time series during recrystallisation which was above the limit of the measurement uncertainties. The curve decays exponentially in the beginning with $E(t) - \langle E \rangle \propto e^{-2.59t}$. The solid red line shows the exponential fit. The dashed black line at 0.1 eV is the quantity substracted from $E(t)$. Below that line, $E(t)$ becomes smaller than two times the measurable limit. Here the decay rate drops slightly before the constant

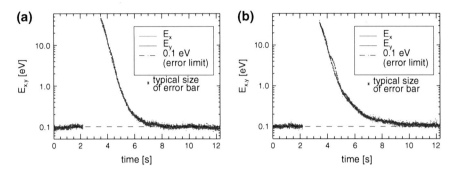

Fig. 7.5 Kinetic particle energies E_x (*red*) and E_y (*blue*) vs. time for experiments rI (**a**) and rII (**b**). The gap between 2 and 4 s corresponds to the time where particles could not be traces sufficiently. The relative errors $\delta E_x / E_x$ and $\delta E_y / E_y$ are $\approx 3.8\%$ in **a** and $\approx 3.3\%$ in **b** (with $\delta E_x, \delta E_y$ are the 1-σ uncertainties from the fit). The *dashed black line* marks 0.1 eV—the limit of the resolvable particle motion, imposed by pixel-noise and tracking uncertainties. Data points close to or below this line are omitted from further interpretation

level is reached at 8 s (end of the x-axis in this figure). The slower decay and spreading of points is presumably due to the influence of the large relative error $\delta E > 50\%$ in $E(t)$ close to 0.1 eV.

Experiment rII $E(t) - \langle E \rangle$ is shown in Fig. 7.6b with an exponential decay $E(t) - \langle E \rangle \propto e^{-2.50t}$ (solid red line). For an interval of ≈ 2 s the decay decreases further to $E(t) - \langle E \rangle \propto e^{-1.25t}$ (solid blue line). Then follows the same phenomenon as in rI for $E(t) < 0.2$ eV (0.2 eV corresponds to all values below the dashed black line, $E(t) - \langle E \rangle < 0.1$ eV) with values spreading broader and decaying slower until the constant level is reached at 10 s. The reason for the slower decay above 0.2 eV ("blue" regime) in experiment rII is due to regions with higher particle mobility. This can be seen in the overlay of all trajectories of the time interval 5.1–9 s in Fig. 7.7. Regions of higher mobility are recognizable as the darker parts in the images, where the trajectories are longer because particles move faster. Two of these regions are marked in Fig. 7.7b (experiment rII) by circles. This leads to broader velocity distributions and thus the seemingly slower decay of $E(t) - \langle E \rangle$ as an artefact of the local phenomenon of "hot spots". In comparison, in experiment rI in Fig. 7.7a, the particle motion is more uniform without localized disturbances.

To identify the "hot spots" as the reason for the slower decay, the kinetic energy was calculated separately for the regions A and B, corresponding to the right (A) and left (B) half of the image in 7.7b. The result is shown in Fig. 7.6c: $E(t) - \langle E \rangle$ from region A (black dots)—the "undisturbed" part—follows the initial exponential decay (red line). The kinetic energy from region B (yellow dots)—containing the marked "hot spots"—is clearly higher and decays slower. The blue line is the same exponetial decay as in Fig. 7.6b, and represents well the average between both.

Conclusion The exponent of the initial decay for both experiments is close to the Epstein drag coefficient $\nu_{Ep} = 2.69$ Hz from the last section. If the smaller $\nu_{Ep} = 2.28$ Hz, calculated with the smaller reflection index, is considered, the system would

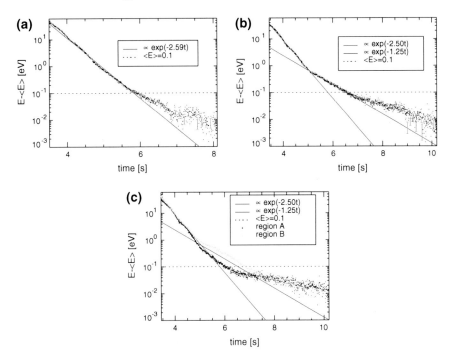

Fig. 7.6 Mean kinetic energy of the particles vs. time calculated as the average of the energies in x and y direction. A ground value $\langle E \rangle$, marked by the *dashed line*, of 0.1 eV has been substracted. The *colored lines* represent exponential fits to the curves (see *inset* for details). The *error bars* (*green*) are representative for all values. **a** Experiment rI, **b** Experiment rII, **c** Experiment rII, with the kinetic energies separatly shown for the regions A (*black dots*) and B (*yellow dots*) of Fig. 7.7

cool down even faster that just due to the friction, but the reflection index in this case was subject to a relatively large error of 0.13 [10]. The coefficient $\delta = 1.442$ originally calculated by Epstein [14] would lead to $\nu_{Ep} = 2.61$ Hz, still larger than the decay rate of the kinetic energy. Considering the uncertainties in δ, it can be well enough concluded that the system initially cools down basically due to Epstein damping, i.e. the damping of the particle motion by friction with neutral gas atoms.

In rII, though, after ≈ 2 s from the beginning of the plot in Fig. 7.6, a process heats particles locally. This decreases the cooling rate even below the Epstein coefficient. The source of increased heat motion was identified as "hot spots" with increased particle mobility in one half of the field of view. The reason for the appearance of this phenomenon is unknown. Since it does not appear in both experiments, it could be a random, local instability arising while particles try to rearrange themselves into the lattice structure. Close to the limit of 0.1 eV, under the influence of the large relative error $\delta E > 50\%$, the energies deviate from the exponential decay and disperse.

Fig. 7.7 Particle trajectories during the time interval [5.1, 9] s after melting. **a** Experiment rI, **b** Experiment rII, circles mark examples for regions with higher particle mobility. The left half of the images corresponds to region B, the right half to region A marked in Fig. 7.1

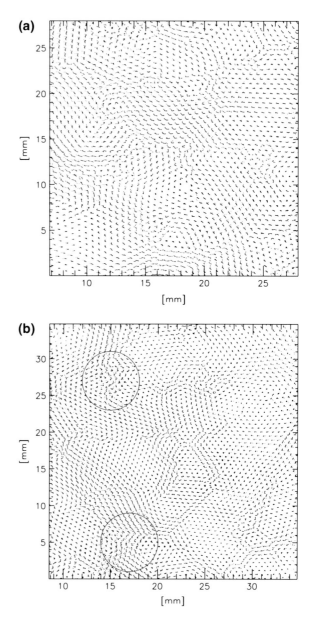

7.5 Global Structural Properties

The pair correlation function $g(r)$ and the bond correlation function $g_6(r)$ are calculated and analyzed to investigate the global system properties with respect to the degree of long range order of the distribution of particles (the translational order) and

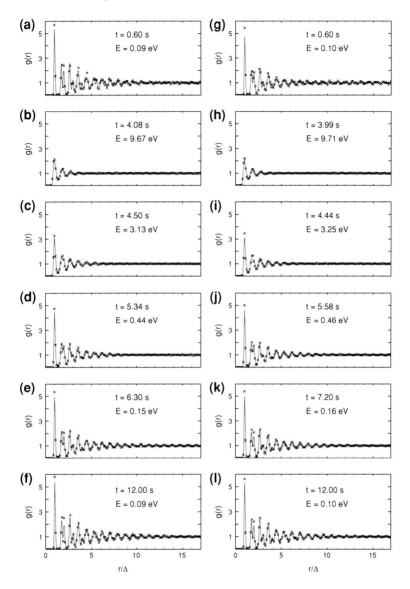

Fig. 7.8 Examples of $g(r)$ during recrystallization. *Solid lines* are fits with the Beresinkii-function. *Left column* (**a–f**): Experiment rI; *right column* (**g–l**): Experiment rII at equal values of the kinetic energy E as rI

of the mutual orientation of nearest neighbor bonds (the orientational order). The definitions and interpretations of both functions were described in detail in Sects. 4.2.2 and 4.2.3 which also gave all nessecary technical information on the calculations and fitting routines.

Of special interest is of course the change of the order during the recrystallization process. Therefore, the quantities derived in the following chapter will be primarily presented in dependence on the progressing time. Later the structural order parameters will be connected with the thermodynamical state of the particles—the kinetic energy.

7.5.1 Translational Order

The pair correlation function $g(r)$ was calculated for each frame up to a distance $r \approx 17\Delta$. Examples for different stages during the recrystallization are shown in Fig. 7.8 for experiments rI (left column) and rII (right column) at levels of equal kinetic energy. $g(r)$ was then fitted by the function given in (4.7) with a variable peak width, as suggested by Beresinskii [15]. The fit is shown as the solid line in Fig. 7.8. Though this model was originally derived for crystalline states with low particle temperatures, it was in good agreement for high particle kinetic energies up to approximately 10 eV.

The exponential fit with fixed peak width from (4.6) worked too, but the goodness-of-fit test produces a larger $\chi^2_{Exp} \approx 1.04 \ldots 2.26 \chi^2_{Ber}$ for experiment rI and $\chi^2_{Exp} \approx 1.00 \ldots 2.25 \chi^2_{Ber}$ for rII for the purely exponential fit in comparison with the Beresinkii function. The ratio of $\chi^2_{exp}/\chi^2_{Ber}$ close to 1 appears at the highest particle energies only. Therefore, the results of the Beresinkii fit were taken.

In experiment rI, the reduced $\chi^2_{Ber,v}$ of the Beresinkii fit was high with 70–90 before melting, and rising up to 60 during recrystallization, with high values in more ordered states. For rII, the value was even higher with nearly 100 in the beginning, but settled at around 25 towards the end of the measurement.

The estimated parameters interparticle distance Δ, initial peak width σ_0 and the translational correlation length ξ are shown in Fig. 7.9. for experiments rI (a–c) and rII (e–f). The fourth fit parameter, connected to the particle density, contains no additional information since it is proportional to Δ^2, but the fitting procedure was more stable when it was allowed to fit this parameter, too.

The fitting routine provides the 1-σ uncertainty in the estimates for each fit parameter. Since the basis for the calculation of the 1-σ uncertainty is the statistical error of each point of $g(r)$, and this error was very small due to the large number of points contributing to the calculation of $g(r)$ as it was already addressed in Sect. 4.2.2, the error bars become very small. Sample error bars representing 1-σ in the plots in Fig. 7.9 give an impression on the quantity of the error. Note that this error bars describe the goodness of the fit, not the measurement errors of physical quantities and that the 4-parameter fit of a rather complex function to $g(r)$ is always at a risk of misinterpretations due to local minima of the parameter set which might make the fit result ambiguous.

Experiment rI The interparticle distance Δ, initialy at 0.592 mm, fluctuates due to the influence of the compressing disturbance of the electric pulse during

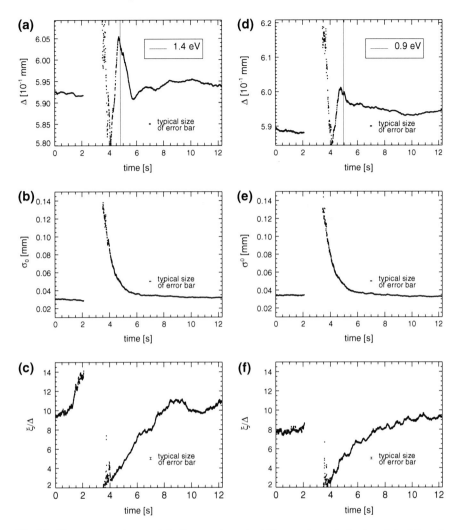

Fig. 7.9 Fit parameters interparticle distance Δ, peak width σ_0 and correlation length ξ of the pair correlation function $g(r)$ for experiments rI (**a–c**) and rII (**d–f**). ξ is normalized by the interparticle distance. The typical size of the (1-σ) uncertainty of the fitted parameters is shown as an example error bar in each plot

an interval of ≈ 2s. It relaxes to a value of 0.594 mm towards the end of the measurement. The peak width σ_0 decays quickly during recrystallization until 0.032 mm close to $\sigma_0 = 0.030$ mm in the beginning. The correlation length ξ starts at $\approx 9.5\Delta$ and exhibits a strong increase up to 14Δ short before the pulse was applied. During recrystallization, it increases from close to zero within 4 s to an average of 10Δ with some fluctuations. The reason the the increase before melting is not clear. From visual inspection of the movies of that time frame it looks like some defects move towards the image edge, and therefore the translational order might spread over a longer range than before. Defects usually influence $g(r)$ by broadening and decreasing the peaks; then the envelope decreases faster.

Experiment rII In contradiction to experiment rI, here Δ settles at a larger value of ≈ 0.595 mm after melting compared to the initial $\Delta < 0.590$ mm. This indicates a shift in the particle density towards a lower value after the recrystallization. A possible reason is a decrease of the system due to the loss of particles during melting, therefore leaving some space for expansion. The peak width σ_0 returns to the initial 0.033–0.034 mm. ξ is in general smaller than in experiment rI: $\xi \approx 8\Delta$ in the initial phase, and it increases to $\approx 9\Delta$ during the recrystallization. There is no sudden increase of ξ before melting as in rI.

Peak Width In crystalline states, the peak width $\sigma_0/\sqrt{2} = \sigma_r$ is a measure of the dispersion of particles at their mean lattice sites, caused by the finite particle temperature, and it is connected to the particle temperature and the Einstein frequency by $\sigma_r = \sqrt{(k_B T)/m}/\Omega_E$ (4.8). In states of high particle temperatures, the mean lattices sites are not clearly defined, if they exist at all. Then σ_r is more likely related to the temperature—or kinetic energy E—by a time scale $1/\Omega$ defined by the underlying dynamic processes, and $\sigma_r = \sqrt{E/m}/\Omega$.

$\sigma_0^2/2$ is plotted versus E in Fig. 7.10. A linear decay $\sigma_r^2 = \sigma_0^2/2 = \sigma_{off}^2 + KE$ was fitted (solid and dashed lines in Fig. 7.10) with

$$
\text{Experiment rI} \quad
\begin{aligned}
\left.\begin{aligned}
\sigma_{off}^2 &= 8.40 \times 10^{-4}\ \text{mm}^2 \\
K &= 2.18 \times 10^{-4}\ \text{mm}^2/\text{eV}
\end{aligned}\right\} &\quad 1.4 < E < 17\,\text{eV} \Rightarrow \Omega = 34.6\,\text{s}^{-1} \\
\left.\begin{aligned}
\sigma_{off}^2 &= 5.39 \times 10^{-4}\ \text{mm}^2 \\
K &= 4.99 \times 10^{-4}\ \text{mm}^2/\text{eV}
\end{aligned}\right\} &\quad E < 1.4\,\text{eV} \Rightarrow \Omega = 22.9\,\text{s}^{-1}
\end{aligned}
$$

$$
\text{Experiment rII} \quad
\begin{aligned}
\left.\begin{aligned}
\sigma_{off}^2 &= 9.21 \times 10^{-4}\ \text{mm}^2 \\
K &= 2.35 \times 10^{-4}\ \text{mm}^2/\text{eV}
\end{aligned}\right\} &\quad 0.9 < E < 13\,\text{eV} \Rightarrow \Omega = 33.3\,\text{s}^{-1} \\
\left.\begin{aligned}
\sigma_{off}^2 &= 5.09 \times 10^{-4}\ \text{mm}^2 \\
K &= 6.24 \times 10^{-4}\ \text{mm}^2/\text{eV}
\end{aligned}\right\} &\quad E < 0.9\,\text{eV} \Rightarrow \Omega = 20.5\,\text{s}^{-1}
\end{aligned}
$$

Remarkable is the change of the slope to a steeper decay at lower energies, and the non-zero offset of σ_r^2.

The offset of σ_r^2 at $E = 0$ is presumably due to the non-ideal lattice structure caused by defects. In an ideal hexagonal crystal at zero temperature, there would be no variation of interparticle spacings, and $\sigma_{off} = 0$ there.

The change of the slope with the particle energy suggests that the typical time scale changes, since $K = 1/(m\Omega^2)$, and the particle mass m is constant. Ω is given in the table above. In the first regime, $\Omega \approx 33\text{–}35\,\text{s}^{-1}$, while in the regime with steeper decay, the frequency is $20\text{–}23\,\text{s}^{-1}$.

Examining other quantities at the time at which the slope changes abruptly, one finds that for both experiments it coincides with a fluctuation of the mean interparticle distance Δ, marked by the vertical solid lines in Fig. 7.9a, d. Apparently the

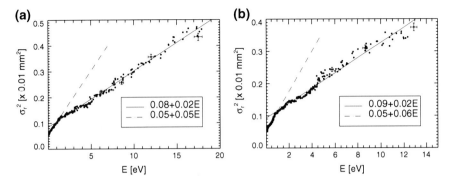

Fig. 7.10 Width of the displacement distribution $\sigma_r^2 = \sigma_0^2/2$ vs. particle kinetic energy E for experiments rI (**a**) and rII (**b**). The *solid* and *dashed lines* are linear fits with slopes given in the legend. The steeper slopes (*dashed*) are fitted for energies $\lesssim 1$ eV. Selected error bars for larger σ_r^2 and E are shown. For values <1 eV, error bars are of the size of the plot symbols. In general, the relative errors of E are of the magnitude 2–3% and constant. The relative errors of σ_r^2 lie in the interval $\{0.5, 1.6\}$% for rI and rII

particle system is expanding at the times to the left of the solid line where the interparticle distance is increasing. This means an increase of the area of the unit cells of particles, causing a decreasing interparticle confinement and therefore an increase in the area particles can visit within their unit cell with respect to their nearest neighbors. Since σ_0 is a measure for that area, it decays slower as long as the system is expanding. This feature does not appear in the kinetic energies defined by the width of the velocity distributions, because the mean motion caused by the pulse only influenced the position of the mean of the distribution, not its width.

It is interesting, that σ_0 does not decay further towards the end of the measurements. If the particle kinetic energy stays at 0.1 eV due to restriction of the measurement uncertainties, other quantities depending on E, which are not affected by the same measurement uncertainties, should decrease until they reach the equivalent of room temperature, since that is the assumed temperature of particles in a complex plasma. The minimum of σ_0 measured from $g(r)$ is approximately 2 times higher than the uncertainty, but $g(r)$ is calculated by counting particle numbers in bins, and the choice of the bin size of $g(r)$ of 2 pixels restricts the resolution of σ_0 to ≈ 0.034 px.

7.5.2 Orientational Order

The bond correlation function $g_6(r)$ measures the average orientation of nearest neighbor bonds separated by the distance r in the crystal (see Sect. 4.2.3). $g_6(r)$ was obtained for each frame up to $r \approx 17\Delta$, and power-law $\propto r^{-\eta_6}$, exponential $\propto e^{-r/\xi_6}$ and linear $\propto c_6 r$ decays were fitted to $g_6(r)$ for $r \geq 3\Delta$. For smaller r the model

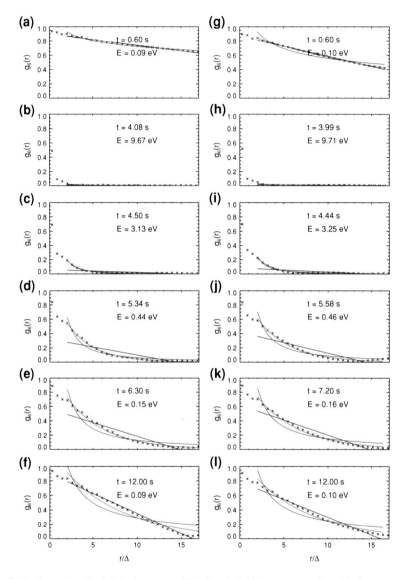

Fig. 7.11 Examples of $g_6(r)$ during recrystallization. *Solid lines* are power law (*blue*), exponential (*red*) and linear (*black*) fits. *Left column* (**a–f**): Experiment rI; *right column* (**g–l**): Experiment rII at equal values of the kinetic energy E as rI

of the long range order does not apply. Some examples for $g_6(r)$ are shown in Fig. 7.11a–f for experiment rI, and Fig. 7.11g–l for rII, taken at the same times as the examples of the pair correlation function in the previous chapter. The three fits are plotted as blue (power law), red (exponential) and black (linear) solid lines.

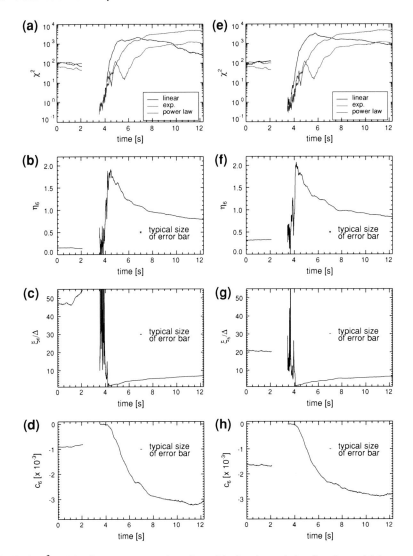

Fig. 7.12 χ_ν^2 and the fit parameters η_6, ξ_6 and c_6 of the bond correlation function $g_6(r)$ for experiments rI (**a–d**) and rII (**e–h**). The smallest χ_ν^2 indicates the best model for the data. ξ_6 is normalized by the interparticle distance Δ. The typical size of the (1-σ) uncertainty of the fitted parameters is shown as an example error bar in each plot

The fits yield the parameters η_6, the bond correlation length ξ_6 and the linear slope c_6, respectively, as they were defined in Sect. 4.2.3. The fit parameters are displayed in Fig. 7.12b–d and f–h. To find the best model for the data, all fits are compared by means of their reduced χ_ν^2 in Fig. 7.12a, e.

The origin of the linear decay, which is not accounted for in the theory, is presumably due to domain forming. This possible explanation was given in Sect. 4.2.3 and is worked out in more detail in Chap. 10. The domain forming was also observed in the presented experiments, as will be shown in a later section.

Experiment rI In the initial state the exponential decay has the lowest χ_ν^2, but with a the correlation length $\xi_6 \approx 47\ldots55\Delta$, which is much larger than the distance of 17Δ up to which $g_6(r)$ has been measured. As is can be seen Fig. 7.11a, the exponential fit (red line) is practically a linear decay. The power law and linear fits seem to be equally good, with $\eta_6 \approx 0.15$ meeting the requirements of the KTHNY theory ($\eta_6 < 0.25$). After melting, between 3.5 and 4.5 s, $g_6(r)$ falls to zero within less than 2–3 interparticle distances, as shown in Fig. 7.11b. The application of any long range order model valid for $r > 3\Delta$ is not appropriate in this regime, and one can safely assume that there is no long range orientational order. From $4.5\ldots5$ s (Fig. 7.11c) on, the power law decay has the smallest χ_ν^2, but $\eta_6 > 1$. The exponential decay with ξ_6 rising from 1.9Δ at 4.5 s to $\approx6\Delta$ at 10 s also seems to be a good model here (Fig. 7.11d, e). It is replaced by the linear decay after 10 s (Fig. 7.11f).

Experiment rII The initial state differs remarkably from experiment rI, in that $\eta_6 \approx 0.32 > 0.25$ and $\xi_6 \approx 20\Delta$ which is of the magnitude of the measured distance. From Fig. 7.11g it is clear that the power law decay does not fit as well as in rI.

The following development of $g_6(r)$ is comparable to rI, with a disordered phase between 3.5 and 4.5 s (Fig. 7.11h). The exponential decay fits well from $\approx4.5\ldots12$ s with $1.9\Delta < \xi_6 < 6.5\Delta$ as can be seen in Fig. 7.11i–l. Though the linear decay has the lowest χ_ν^2 for times after 11 s, the line seems not to fit as well as it does at the end of experiment rI.

7.5.3 Conclusion: Global Order

The long range translational order is high in the initial state, and $g(r)$ has pronounced peaks for large r, as can be seen in Fig. 7.8a, g. This state is also reached again at the end of the time series. The correlation length is high with $\xi > 8\Delta$ in the ordered states at the beginning and end. Short after melting long range order is lost and $\xi \approx 2\Delta$, but it rises fast towards the initial value. The model which was fitted to $g(r)$ did include an exponential decaying envelope, but also an algebraic decay in the factor $\sqrt{\ln(r/r_0)}$ (4.7). The exponential decay seems to be prominent, and no purely power law decays could be fitted to $g(r)$. According to the KTHNY theory, this indicates a liquid or hexatic state, but this interpretation contradicts with the shape of $g(r)$. Since the fit model is modified, a direct comparison to the theory might be difficult.

Further the split of the second peak can be taken as an indicator of a crystalline state. It clearly appears in the initial state, and short after melting the second peak was found to split at approximately 1 eV (between Fig. 7.8c, d (rI) and Fig. 7.8i, j (rII)). Another interesting feature is the change of the linear slope of σ_0^2 in dependence of the particle kinetic energy E at 1.4 eV (rI) and 0.9 eV (rII), which can most likely be attributed to the remaining disturbance of the system by the propagation of the shock induced by the electric pulse.

The orientational correlation function is neither constant nor does it show a pronounced power law decay. The power law can be fitted short after melting in a time window of about 1 s, but with a very high exponent > 1.5, which contradicts the predictions of $\eta_6 < 0.25$ in the hexatic phase in which a power law decay appears according to KTHNY. After that short phase, exponential decays are found with a correlation length ξ_6 increasing up to $< 7\Delta$, which is smaller than the translational correlation length. Towards the end of the measurement, linear decays are found, as in the initial state. It is worth to note that the grain boundary theory of melting suggested an exponential decay of g_6 in the case that no bound states exist between grain boundaries, and a power law decay if there are bound states, because then the net Burgers vector would be zero, and long range order could persist [16]. The linear decay was explained as a possible effect of domain forming (see also Chap. 10). Then the regime of linear decay could be seen as high ordered states with respect to the orientational order on a small length scale, but not on long ranges. Those states are reached 8–9 s after melting, which is more than 2 s after the translational long range order was restored.

In conclusion, the system initially is in a state with long range translational order, and some intermediate range orientational order, which are both destroyed by the melting process. During the recrystallization, long range translational order is restored fast within approximately 5 s after melting, while the long range orientational order is not restored to its initial state until 3 s later, defined by the moment of occurring linear decay of $g_6(r)$. Still, from Fig. 7.11f, l one can see that $g_6(r)$ decays much faster at the end of the measurement than in the initial state. The apparent absence of long range orientational order with respect to common theories will be investigated qualitatively from the point of view of a localized order analysis in the next Sect. 7.6.3. In any case, the behavior of the cooling complex plasma seem to deviate from the predictions of the KTHNY theory, at least from the point of view of the correlation function analysis: The translational order should have been restored at lower temperatures than the orientational order, which is clearly not the case. Also the occurrence of an intermediate hexatic phase is questionable, since the expected power-law decay with $\eta < 0.25$ of the orientational correlation function does not appear.

7.6 Local Structural Properties

As a measure for the degree of short range order the bond order parameter and the average defect fractions are calculated. Those quantities are defined locally and measure the goodness of the structure of the lattice directly at the respective particle positions. The defects are further evaluated as to their distribution and arrangement across the lattice structure.

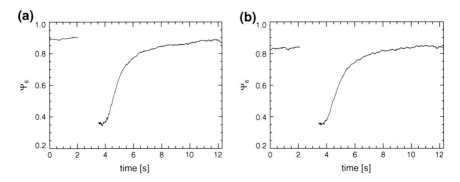

Fig. 7.13 Bond order parameter Ψ_6

7.6.1 Bond Order Parameter

The average bond order parameter Ψ_6 shown in Fig. 7.13 was obtained by calculating $\Psi_{6,k}$ from (4.10) for each unit cell—or particle—k and then averaging its modulus over all cells in one frame. Ψ_6 is a measure for the local order in the system: The closer it is to one, the closer the cells are to an ideal hexagon in average. Lattice sites with nearest neighbor bond angles deviating from 60° will decrease Ψ_6. The same accounts for lattice sites with a number of nearest neighbors other than six.

Experiment rI has a better local order with $\Psi_6 = 0.89$ before and 0.87 after melting, while in experiment rII it was 0.83 and 0.84, respectively. The minimum was $\Psi_6 \approx 0.34$ in both cases. This minimum value seem not to be coincidental, but was also found in another, completely unrelated experiment [17].

The use of the bond order parameter as an averaged value gives a first impression on the degree of short range order in the particle system, but not on its distribution across the ensemble. After examining the defects in the next section, a more qualitative analysis will illustrate the spatial distribution of short range order.

7.6.2 Defect Fractions

Any lattice site with a number of neighbors deviating from 6 is a defect lattice site, or disclination. Most common defects in a two dimensional hexagonal lattice are 5- and 7-fold lattice sites. Are two defects located next to each other in such a way that a non-zero Burgers vector emerges, this pair forms a dislocation.

The number of nearest neighbors of each particle is obtained by performing Delauney triangulations for each frame, yielding the fraction N_k/N of N_k disclinations with k nearest neighbors. N is the total number of particles in a frame.

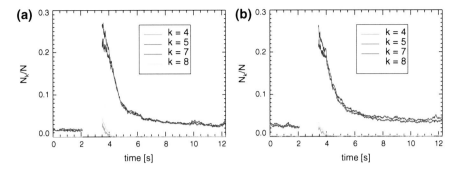

Fig. 7.14 Defect fractions for experiments **a** rI and **b** rII. The colors *green, red, blue* and *yellow* mark the fraction of particles with the number of nearest neighbors 4, 5, 7 and 8, respectively

N_k/N is shown in Fig. 7.14a, b for experiments rI and rII, for $k = 4, 5, 7, 8$. The fractions for $k = 4, 8$ are interesting in the short time directly after melting only, while 5- and 7-folds prevail at most times.

Experiment rI In the initial state, the defect fractions are very low, with 1.8% 5-fold and 1.5% 7-fold lattice sites. Shortly after melting, the fraction of pentagons decreases linear and very fast from 26% at $t = 3.5$ to 8% at 5 s, that of heptagons from 21% down to 7%. The decay rate then slows down and saturations are reached in the last seconds of the measurement with $N_5/N = 3\%$ and $N_7/N = 2.8\%$.

The fractions of particles with 4 or 8 nearest neighbors is lower than 0.03% in the beginning and drop simultaneously to less than 0.06% after $t = 4$ s from their maxima of 2.7 and 4% at $t = 3.5$ s.

Experiment rII The development of the defect fractions with time are practically the same as in rI. 5-fold lattice sites start at 3%, and then drop from 26% at $t = 3.44$ s to 6.8% at $t = 5.5$ s, while the 7-folds go from 21 to 6.3% in the same time interval. The final values are 4 and 3.5%, respectively.

For $k = 4, 8$, the initial values are higher with $N_{4,8}/N \approx 0.1\%$, as are the final fractions with $N_4/N = 0.04\%$ and $N_8/N = 0.18\%$.

Defect Condensation Parameter The defect fraction itself gives no insight in the arrangement of defects within the system. For this purpose the defect condensation parameter S, introduced in [6], is calculated. S is the average number of nearest neighbors of a defect lattice site that are defects themselves. In case of isolated disclinations, the defect condensation parameter would become zero, while it would be one if all defects would be organized in dislocations with exactly one defect neighbor. In case of defect string formation, most defects would have two defect neighbors, and S would become two. Higher numbers are then reached for larger defect clusters, i.e. more than two neighbors of a defect are defects, too. One has to be careful, though, because S is not unique. Equal contributions of isolated defects (0) and defect strings (2) might result in an average $S = 1$.

Figure 7.15a, b show the defect condensation parameter for rI and rII versus time. For clarification of the situation, Fig. 7.15c, d show the fraction of defects with a

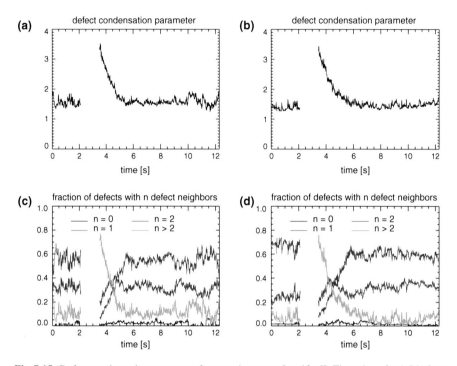

Fig. 7.15 Defect condensation parameter for experiments **a** rI and **b** rII. The value of ≈1.5 before and after recrystallization indicates the forming of defect strings as a significant feature in the arrangement of disclinations, as can be verified by the fraction of defects with *n* defect neighbors, plotted in **c** for rI and **d** for rII. The *colored lines* are the fraction of defects with one (*red*), two (*blue*), more than two (*green*) and no (*black*) adjacent defect

particular number *n* of defect neighbors $n = 0, 1, 2$ and $n > 2$. In the following, values are given for rI, and rII in brackets: *S* stays at 1.5 (1.4) in the initial phase, with 56% (67.5%) from dislocations, 33% (24.1%) from defect chains and 10% (6.7%) from larger defect clusters. Isolated disclinations play no significant role with less than 1% (1.7%).

In the early recrystallization stage, defects seem to be organized in larger clusters, starting at an average of approximately 3.5 defect neighbors. *S* then decreases quickly within less than 2 s to the initial situation of 1.5 defect neighbors, again with the percental contributions of dislocations, strings and clusters as in the initial state. Obviously there is no phase dominated by isolated disclinations. The ground state of the systems consists mainly of dislocations and defect strings as a significant feature. Defect clusters are important in the high temperature phases only. There even isolated dislocations are rare, but for large numbers of defects in disordered states the probability to find more than one defect in the vicinity of another is also high.

7.6.3 Spatial Distribution of Defects and Local Order

Both, defects and the bond order parameter are in contrast to the correlation functions locally defined quantities. In their presentation as ensemble averages above it is not possible to deduce on their distribution across the system. The defect condensation parameter gave a first impression of the arrangement of defects.

Another, more qualitative representation of the data is presented in Figs. 7.16 and 7.17. It shows greyscale maps of $|\Psi_{6,k}|$ for selected frames, with brighter shades of grey corresponding to unit cells with higher order. The arrows represent the vector field of $\Psi_{6,k}$ introduced in Sect. 4.2.4. It is a measure for the orientation of unit cells as to the x-axis of the images, mapping the angles between nearest neighbor bonds $[0, \pi/6] \rightarrow [0, \pi]$ and $[\pi/6, \pi/3] \rightarrow [-\pi, 0]$. Arrows pointing in opposite directions therefore mean that the unit cells are rotated by 30% as to each other, which is the maximum of the difference in the orientation. The red and blue dots mark the locations of 5- and 7-fold defects. The location of bond orientation jumps indicated by the vector field is clearly correlated with the lines of defect locations.

Experiment rI In the initial crystal (Fig. 7.16a) the cells are close to the ideal hexagonal state and mostly identically oriented (practically no abrupt differences of $\arg(\Psi_{6,k})$ are found). After shock-melting, a disordered liquid-like state forms (Fig. 7.16b). Crystallization proceeds first to a system of small ordered 'crystallites' with arbitrary orientations separated by strings of defects. As the system cools down, these crystallites grow and merge with neighboring regions (Fig. 7.16c, d), causing the bonds to tilt to the (single) orientation of the growing region. A metastable state is reached which is characterized by highly ordered adjoined crystalline domains (Fig. 7.16e). Since $\arg(\Psi_{6,k})$ changes abruptly across the domain boundaries—even to directions with opposite signs for adjoining domains—long range orientational order cannot be found in $g_6(r)$ in contrast to the initial state. Locally, the quality of the hexagonal cells is good within larger regions, and deteriorates towards defect locations, where the lattice structure becomes distorted by the defect lattice sites.

Experiment rII The frames shown for experiment rII are chosen for comparable kinetic energies of the particles as for rI. The system is less uniform in the beginning compared to rI (Fig. 7.17a). Unit cell orientations differ between domains, and there are disturbances in the upper left and lower right image edges. The quality of hexagonal cell is good only in the central region. Again, strings of defect pairs can be seen between the domains of different orientation. After melting, a disordered state like in experiment rI appears (Fig. 7.17b), followed by the rapid cooling phase (Fig. 7.17c, d). Here, the crystallite forming is not so strong pronounced as in rI, and regions with considerably higher defect fractions appear, especially in the upper left half of the image. This coincides with the observations of "hot spots" in the analysis of the kinetic energy. The final state (Fig. 7.17e) consists of domains with different unit cell orientation separated by defect strings, but with a visibly higher defect fraction, as it was found earlier (see Fig. 7.14).

Defect Strings Apparently from Figs. 7.16 and 7.17, defects seem to form strings consisting of alternating 5- and 7-fold lattice sites. These strings separate domains

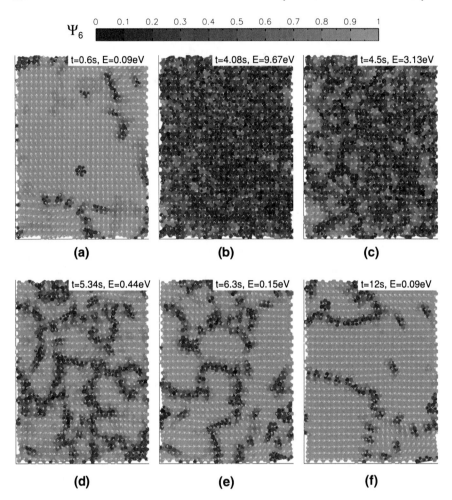

Fig. 7.16 Experiment rI: Development of the local order parameter $\left|\Psi_{6,k}\right|$ (*grey-shaded unit cells*) and the orientation of unit cells, represented by the argument of $\Psi_{6,k}$ plotted as arrays, during recrystallization. The *red* and *blue dots* mark 5- and 7-fold defect lattice sites. **a** $t = 0.6\,\text{s}, E = 0.09\,\text{eV}$. **b** $t = 4.08\,\text{s}, E = 9.67\,\text{eV}$. **c** $t = 4.5\,\text{s}, E = 3.13\,\text{eV}$. **d** $t = 5.34\,\text{s}, E = 0.44\,\text{eV}$. **e** $t = 6.3\,\text{s}, E = 0.15\,\text{eV}$. **f** $t = 12\,\text{s}, E = 0.09\,\text{eV}$

of high local order from each other, as illustrated by the high bond order parameter of the single unit cells. From the defect condensation parameter it is known, that more than 50% of defects are organized in dislocations, and not in defect strings. The above figures make it clear, though, that the dislocations might not be connected with each other directly, but nonetheless form chain-like structures across a considerable distance. It is not possible to show here a long time series of above maps, but a movie of the series of maps of each frames shows that the chains are also stable over a longer time [18].

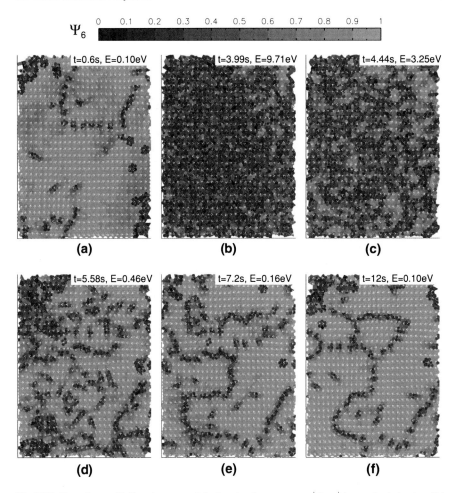

Fig. 7.17 Experiment rII: Development of the local order parameter $|\Psi_{6,k}|$ (*grey-shaded unit cells*) and the orientation of unit cells, represented by the argument of $\Psi_{6,k}$ plotted as arrays, during recrystallization. The *red* and *blue dots* mark 5- and 7-fold defect lattice sites. **a** $t = 0.6$ s, $E = 0.10$ eV. **b** $t = 3.99$ s, $E = 9.71$ eV. **c** $t = 4.44$ s, $E = 3.25$ eV. **d** $t = 5.58$ s, $E = 0.46$ eV. **e** $t = 7.2$ s, $E = 0.16$ eV. **f** $t = 12$ s, $E = 0.10$ eV

An attempt to identify the strings of defects by assigning pairs of 5- and 7-folds is shown in Figs. 7.18 and 7.19 for the same time steps as in Figs. 7.16 and 7.17. The unit cells are as before colored according to their local bond order parameter. The arrows between 5- and 7-folds indicate a dislocation. The assignation procedure was described in Sect. 4.2.1. It searches all lattice sites adjacent to a defect for another defect successively until no more defects are found. That ensures that chains and clusters are identified, additionally to isolated disclinations or dislocations.

Experiment rI Except on the image edges, no isolated disclinations can be found in the initial state (Fig. 7.18a). Instead the defects are arranged in either

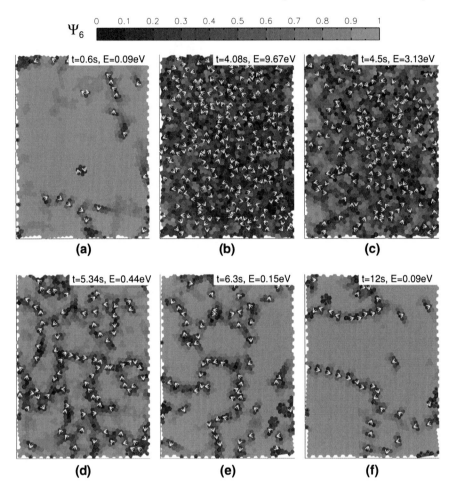

Fig. 7.18 Experiment rI: Development of defect chains during recrystallization. *Arrows* connect one 5-fold (*red dot*) with an adjacent 7-fold (*blue dot*) defect. Unit cells of particles are *grey-shaded* according to their bond-order parameter. **a** $t = 0.6$ s, $E = 0.09$ eV. **b** $t = 4.08$ s, $E = 9.67$ eV. **c** $t = 4.5$ s, $E = 3.13$ eV. **d** $t = 5.34$ s, $E = 0.44$ eV. **e** $t = 6.3$ s, $E = 0.15$ eV. **f** $t = 12$ s, $E = 0.09$ eV

isolated dislocations, or chains of alternate 5- and 7-fold lattice sites. But the free dislocations (lower half of the image) are not randomly distributed, but form a chain-like structure, separated only by a few hexagonal cells. Directly after melting in the disordered state, assignation of defect pairs is more or less random (Fig. 7.18b, c). A lot of disclinations are not assigned in that regime. Later more and more dislocations are uniquely identified, and strings of defect pairs are formed again (Fig. 7.18d–f). The interruption of defect chains by 6-fold lattice sites explains the outcome of the defect condensation parameter, which showed that isolated dislocations dominate the system at most times.

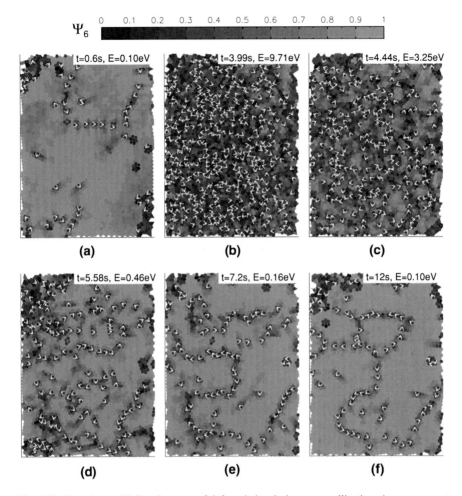

Fig. 7.19 Experiment rII: Development of defect chains during recrystallization. *Arrows* connect one 5-fold (*red dot*) with an adjacent 7-fold (*blue dot*) defect. Unit cells of particles are *grey-shaded* according to their bond-order parameter. **a** $t = 0.6$ s, $E = 0.10$ eV. **b** $t = 3.99$ s, $E = 9.71$ eV. **c** $t = 4.44$ s, $E = 3.25$ eV. **d** $t = 5.58$ s, $E = 0.46$ eV. **e** $t = 7.2$ s, $E = 0.16$ eV. **f** $t = 12$ s, $E = 0.10$ eV

Experiment rII The interpretation is practically the same as in rI, shown in Fig. 7.19a–f. There is a visible higher defect fraction, and a few isolated disclinations can be seen. All other defects are found in pairs or chains, and chain-like structures with dislocation pairs alternating with 6-fold lattice sites.

7.6.4 Conclusion: Local Order

The local order given by the bond order parameter is high before melting with an average $\Psi_6 \in [0.8, 0.9]$. This state is reached again within ≈ 2.5 s after melting

Fig. 7.20 Total Burgers
vector normalized by the
number of dislocations vs.
time for experiment rI

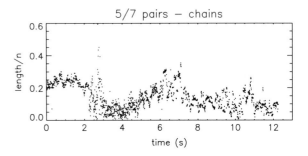

(Sect. 7.6.1). Also the defect fraction quickly return to low values shortly after the system was melted (Sect. 7.6.2). Clearly the system initially was in a highly ordered state, where most defects were organized in dislocations, as indicated by the defect condensation parameter. The dislocations formed strings or string-like structures interrupted by hexagonal unit cells. The quality of hexagons was diminished in the vicinity of the dislocations, where the crystal had to arrange its structure due to the distortion (Sect. 7.6.3). Shortly after melting, the defect fraction was high and disclinations were randomly distributed. Defects were mostly in larger clusters. The system was in a high disordered state, but quickly changed to a regime of domain forming, with domains characterized by a uniform, high bond order parameter within the domain. Domains were separated by defect strings, and they were growing as the system cooled down, until a state similar to the initial was obtained, and few strings of dislocations prevailed. The higher percentage of free dislocations than strings of dislocations in the defect condensation parameter is caused by the interruption of the strings by non-defect lattice sites, but the geometrical arrangement could clearly be seen in Figs. 7.16 and 7.17.

The strings were identified in Sect. 7.6.3, and it would now be possible to calculate a total Burgers vector, since the vector between the two components of a dislocation is perpendicular to its Burgers vector. If the total Burgers vector is 0, the system would have bound states of dislocations, and long range order could exist [19, 16]. The total Burgers vector has been calculated for experiment rI after assigning the defect pairs. Its length, normalized by the number of dislocations, is plotted in Fig. 7.20. It is higher in the initial state with 0.2, then it falls close to 0 after melting. Until the end of the measurement it fluctuates between 0 and 0.2. It can not be decided from this quantity, if a zero value comes from bound dislocations or from random extinction of vectors. Since only a section of the whole particle system is seen in the field of view of the camera, it is impossible to obtain a meaningful net Burgers vector. Even if the whole system would be analyzed, from the net Burgers vector alone the existence of bound dislocation pairs, as proposed by the KTHNY theory, can not be distinguished from bound states of grain boundaries, as mentioned in the grain boundary theory.

Especially a binding of free dislocations into pairs with opposite Burgers vector, as it is expected in the KTHNY theory of dislocation mediated melting, could not be observed. The forming of domains with different lattice orientation, as is was found qualitatively here, is the reason for the absence of long range orientational order in $g_6(r)$ and its linear decay, confirmed by the results in Chap. 10. Orientational long range order can only exist within a domain, as soon as a boundary is crossed, the order is destroyed.

A computer simulation was performed with 816 particles and identical conditions as in the above experiments. The particle cluster was heated to 10 eV and then left alone to crystallize. It was found that the particle temperature had the same exponential decay as found in the experimental data, and the formation of a crystallite, separated from the bulk by a circular grain boundary, was observed [9].

The observed tendency of dislocations to form grain boundaries was found in another experiment with two-dimensional plasma crystals at constant external conditions: Pairs of dislocations where continuously created by applying shear stress to the 2D crystal [20]. The pair then split into two free dislocations, which moved away from the generation area in opposite directions, and sedimented into parallel grain boundaries.

The impact of the grain boundaries will be investigated closer in the next chapter.

7.7 Boundary Energy

From the previous results it is obvious that defects are organized in chain-like structures, and these structures define boundaries between regions of different unit cell orientation. An estimate for the energy stored in those boundaries can be obtained by calculating the average energy $\langle E_{5,7} \rangle$ of all 5- and 7-fold defects in one frame, and the average "bulk" energy $\langle E_6 \rangle$ stored in the hexagonal, non-defect lattice sites. Then the ratio $\langle E_{5,7} \rangle / \langle E_6 \rangle$ gives an estimate of the average excess energy of a defect lattice site located in a boundary.

Due to the high uncertainties in the velocities especially at the beginning and end of the time series, instead of directly calculating the kinetic energy or temperature, the mean square displacement (MSD) of particles, $\langle r^2 \rangle_k$, is calculated for the bulk ($k = 6$) and the boundaries ($k = 5, 7$) separately in each frame, as a measure for the mobility of the particles. r is the modulus of the displacement vector calculated in the local coordinate system with (4.17), and r_x, r_y are its components:

$$\langle r^2 \rangle_k = \frac{1}{N_k} \sum_{i=1}^{N_k} r_{i,x}^2 + r_{i,y}^2 \qquad (7.4)$$

with $k = 6$ or $k = 5, 7$ and the sum going over all N_6 6-fold particles in one frame for the bulk, $\langle r^2 \rangle_6$, and over all defect particles N_5 and N_7 for the boundary, $\langle r^2 \rangle_{5,7}$.

The behavior of $\langle r^2 \rangle_k$ should represent the behavior of the kinetic energy of the particles in a qualitative way. However, the local coordinate system should not be understood in the context of an equilibrium mean lattice site, since this can only

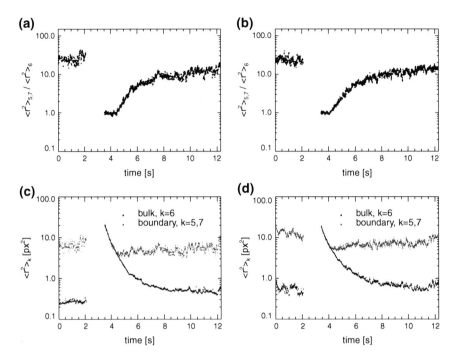

Fig. 7.21 Ratio of the mean squared displacement (MSD) of particles in boundaries (defect chains) to the mean squared displacement of particles in the bulk (6-fold lattice sites). **a** Experiment rI. **b** Experiment rII. MSD of particles in boundaries (*red dots*) and MSD of particles in the bulk (*black dots*): **c** Experiment rI. **d** Experiment rII

be defined clearly in a crystalline state. The above summation rather represents the average mobility of particles within the area defined by their nearest neighbors, which is a measure for the kinetic energy. $\langle r^2 \rangle_{5,7}/\langle r^2 \rangle_6$ is shown in Fig. 7.21a, b for experiment rI and rII. Also shown are the components $\langle r^2 \rangle_{5,7}$ and $\langle r^2 \rangle_6$ in Fig. 7.21c, d.

Both experiments rI and rII show the same behavior. Shortly after melting at 4 s the average energy in boundary and bulk is equal. This corresponds to a time where the defect fraction is high, and defects are mostly randomly distributed. The system cools down due to damping of the particle motion (the decay found in Sect. 7.4), but within less than one second the ratio increases and from Fig. 7.21c, d it is clear that the average energy in the boundaries stays constant, while the bulk energy continues to decrease. For comparison with the arrangement of defects in the system, the increase of the energy ratio appears at a time between Fig. 7.16b, c (rI) or Fig. 7.17b, c (rII). Subsequently the boundaries start to form between domains and the domains grow. During all this, the average energy in the boundaries seems to be constant and higher by a factor of 10 (rI and rII) than the energy stored in the bulk. In the initial state before melting, this ratio was even higher with ≈ 25.

Table 7.3 Exponents β from power law fits to the correlation lengths, defect fractions and bond order parameter

	β	
	Experiment rI	Experiment rII
ξ	−0.28	−0.26
ξ_6	−0.33	−0.32
N_5/N	+0.34	+0.34
N_7/N	+0.33	+0.35
$(N_5 + N_7)/N$	+0.34	+0.35
Ψ_6	−0.16	−0.16

The separation of domains with different lattice orientation by grain boundaries with a high energy stored in the boundaries was also found in experiments with 3D complex plasmas [21]. There the boundaries seemed to be molten as to their energy, but optically showed a structural order.

Apparently, the particles located in the grain boundaries have a considerable higher mobility (higher kinetic energy) than particles in the bulk. This has to have some impact on the behavior of the complete system, which will be investigated closer when the nature of the phase transition is considered in Sect. 7.9.

7.8 Connection Between Structure and Kinetic Energy

The dependence of structural properties of the particle system during recrystallization on the particle kinetic energy is investigated. From this connection between fundamental dynamical properties and a thermodynamical quantity, insights into the nature of the phase transition are obtained.

7.8.1 Power Law Scaling

The parameters describing the long range and the local order in the particle system, namely the translational and orientational correlation lengths ξ and ξ_6, the fraction of defects N_k/N ($k = 5, 7$) and the average bond order parameter Ψ_6, are plotted versus the average kinetic particle energy E in double-logarithmic plots, shown in Fig. 7.22 for experiment rI, and Fig. 7.23 for experiment rII. All quantities exhibit a power law dependence on the average particle kinetic energy E. Fits cE^β with a constant c and the exponent β were performed for all quantities, and plotted as black lines in the figures.

Table 7.3 lists all exponents β. Additionally is shows β for a fit to the defect fraction of all defects, $(N_5 + N_7)/N$, which is not shown in the figures.

The power law dependencies on E of all presented quantities suggest that the system is scale free during recrystallization. Especially the proposed exponential

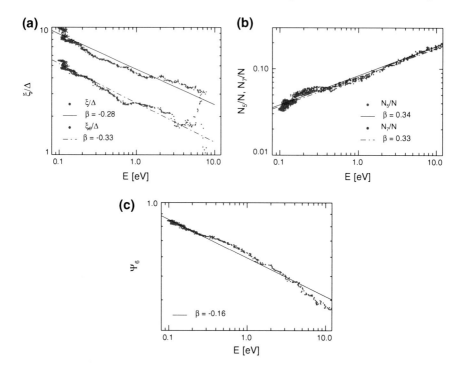

Fig. 7.22 Experiment rl: Dependence on the kinetic energy E of **a** translational and orientational correlation length ξ/Δ (*red*) and ξ_6/Δ (*blue*), **b** defect fractions N_5/N (*red*) and N_7/N (*blue*), and **c** bond order parameter Ψ_6. The *black lines* are power law fits $\propto E^\beta$

dependencies of ξ and ξ_6 on the energy (or temperature) from the KTHNY theory of melting (Sect. 6.1.2) can not be confirmed.

Interesting would be the dependence of the coupling parameter Γ, estimated with the relation found in Chap. 5, on the kinetic energy E. Generally, $\Gamma \propto 1/E$ would be expected by definition of Γ. But the derived relation might not be correct in the case of high energies, because the idea of an oscillation of particles around their mean lattice site is not applicable. The mean squared displacement from the last chapter could be used as a substitute for σ_r, but it is questionable if the value calculated with it is comparable to known values of Γ. The same argumentation can be applied for the Lindemann criterion, which basically states a critical value at the melting point $= 2/\Gamma$.

The next chapter will introduce a possible theoretical explanation for the power law decays, based on the theories of Frenkel (Sect. 6.4, [22]).

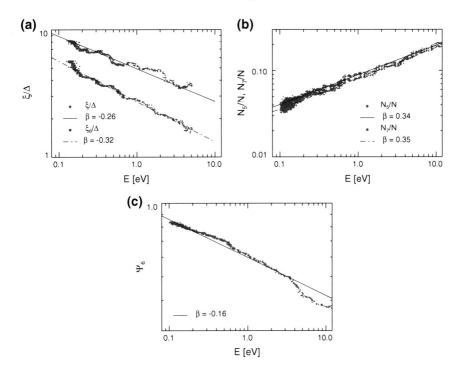

Fig. 7.23 Experiment rII: Dependence on the kinetic energy E of **a** translational and orientational correlation length ξ/Δ (*red*) and ξ_6/Δ (*blue*), **b** defect fractions N_5/N (*red*) and N_7/N (*blue*) and **c** bond order parameter Ψ_6. The *black lines* are power law fits $\propto E^\beta$

7.8.2 Theory for the Crystallization of a 2D Complex Plasma

The surprising result of power law dependencies of structural quantities from the last chapter might be connected to the domain structure which developed during the recrystallization. The Arrhenius law for the dependence of the defect fractions on the particle temperatures in (6.7) was applicable for systems in thermodynamical equilibrium. The recrystallization experiments presented in the last chapter involved the rapid cooling of a two-dimensional particle system, which is basically a non-equilibrium situation.

Based on the work of Frenkel [22], a theoretical model is developed to explain this behavior for a two-dimensional system [23].

A 2D system at a temperature T consisting of N particles is divided into $z = N/\bar{N}_d$ homogenous domains each containing $\bar{N}_d$ particles on average. The domains are separated by boundaries consisting of dislocations—pairs of 5- and 7-fold defects. There are no correlations of the structural order between the domains. For the moment only averages are considered in the following. The mean domain radius $\bar{r}$ is derived from the domain area which is equal to the sum of all unit cells around the particles. The interparticle separation is Δ.

$$\pi \bar{r}^2 = \frac{1}{4}\bar{N}_d \pi \Delta^2 \quad \Rightarrow \quad \bar{r} = \sqrt{\bar{N}_d}\frac{\Delta}{2} \tag{7.5}$$

The additional energy $\bar{E}$ of the domain boundaries is

$$\bar{E} = 2\pi \bar{r}\sigma z = \pi \Delta \sigma \sqrt{N z} \tag{7.6}$$

with the surface tension σ.

The available number of microstates P accessible to the system can be obtained by a calculation of the number of possibilities to distribute N distinguishable particles on z regions with $\bar{N}_d$ particles in each region:

The number of possibilities P_1 to choose $\bar{N}_d$ distinguishable particles out of N is

$$P_1 = \binom{N}{\bar{N}_d} = \frac{N!}{\bar{N}_d!(N - \bar{N}_d)!} \tag{7.7}$$

Now again $\bar{N}_d$ particles are chosen from the remaining $N - \bar{N}_d$ with the number of possibilities P_2:

$$P_2 = \binom{N - \bar{N}_d}{\bar{N}_d} = \frac{(N - \bar{N}_d)!}{\bar{N}_d!(N - 2\bar{N}_d)!} \tag{7.8}$$

This is repeated until all z regions are filled. Since the events P_i, $i \in [1, z]$, are independent from each other, the total number of possibilities P to arrange the N particles is the product of all P_i:

$$
\begin{aligned}
P &= \prod_{i=1}^{z}\binom{N - (i - 1)\bar{N}_d}{\bar{N}_d} \\
&= \frac{N!}{\bar{N}_d!(N - \bar{N}_d)!} \cdot \frac{(N - \bar{N}_d)!}{\bar{N}_d!(N - 2\bar{N}_d)!} \cdot \frac{(N - 2\bar{N}_d)!}{\bar{N}_d!(N - 3\bar{N}_d)!} \cdots \frac{(N - (z - 1)\bar{N}_d)!}{\bar{N}_d!(N - z\bar{N}_d)!} \\
&= \underbrace{\frac{N!}{\bar{N}_d! \cdot \bar{N}_d! \cdots \bar{N}_d!}}_{z times} \cdot \underbrace{\frac{1}{(N - z\bar{N}_d)!}}_{=0!=1} = \frac{N!}{[\bar{N}_d!]^z}
\end{aligned}
$$

From this follows for the entropy of the system:

$$S = k_B \ln P \tag{7.9}$$

with the Boltzmann constant k_B. Using Stirling's approximation, $\ln(N!) = N \ln(N) - N$ for large N (for $N > 1000$ the relative error of the approximation is $< 1\%$), with N, $\bar{N}_d$ sufficiently large and $\bar{N}_d = N/z$, the term $\ln P$ can be approximated:

$$
\begin{aligned}
\ln P &= \ln\left(\frac{N!}{[\bar{N}_d!]^z}\right) = \ln(N!) - z \ln[(N/z)!] \\
&\approx N \ln(N) - N - z\frac{N}{z}[\ln(N) - \ln(z)] + z\frac{N}{z} = N \ln(z) \tag{7.10}
\end{aligned}
$$

Therefore we get for the entropy and the free energy F for the system at temperature T:

$$S = Nk_B \ln z \tag{7.11}$$

$$F = \bar{E} - TS = \pi \Delta\sigma \sqrt{Nz} - NK_BT \ln z \tag{7.12}$$

At any instant during the cooling process we may assume that the free energy is independent of the distribution of domains in the system and that parameters such as temperature and pressure are the same within all domains. Then

$$\frac{\partial F}{\partial z} = 0 \quad \Rightarrow \quad 0.5\pi \Delta\sigma \sqrt{N/z} - Nk_BT/z = 0$$

$$\Rightarrow \quad z = \left(\frac{2k_BT}{\pi \Delta\sigma}\right)^2 N \tag{7.13}$$

Equation 7.13 gives a relationship between the number of domains and the temperature. Note that z does not depend on the average number of particles in a domain $\bar{N}_d$.

Now the fractal nature of the domains is introduced as a hypothesis. The following equation then connects the domain area to the length of its boundary:

$$\bar{N}_d \Delta^2 B = [\Delta \bar{N}_S]^{1+\alpha} \tag{7.14}$$

with B a constant and $\bar{N}_S$ the average number of particles in the domain wall. B and α are dependent on the shape of the domain, i.e.

circular domain	$B = \pi^2,$	$\alpha = 1$
compact domain ($\bar{N}_d \rightarrow \bar{N}_S$)	$B \rightarrow 1/\Delta,$	$\alpha \rightarrow 0$
long narrow domain	$B \rightarrow 2/\Delta,$	$\alpha \rightarrow 0$

For real (fractal) 2D systems, $0 < \alpha < 1$ is expected.

With $\bar{N}_d = N/z$ and z from (7.13), (7.14) yields for the total number of particles located in domain walls, $N_T \equiv z\bar{N}_S$:

$$\frac{N_T}{N} = B^{1/(1+\alpha)} \Delta^{(1-\alpha)/(1+\alpha)} \left(\frac{2k_BT}{\pi \Delta\sigma}\right)^{(2\alpha)/(1+\alpha)} \tag{7.15}$$

In the above calculation it is assumed that the surface tension σ is constant, especially that there is no temperature dependence of σ.

7.8.3 Comparison with the Measurements

The experimentally found power law decay of the defect fractions is now compared with (7.15), with the kinetic energy E representing the particle temperature k_BT.

The temperature dependence of Δ is now assumed to be weak enough to be neglected: the total fluctuation of Δ during recrystallization is 4.2% in experiment rI and 2.7% in experiment rII (% of the average value) as can be seen in Fig. 7.9a, b. Assuming that all defects are located in boundaries, so that $(N_5 + N_7)/N = N_T/N$ and the surface tension σ does not depend on E, and further assuming that the domains and boundary lengths have in fact a fractal dependence, it follows from the performed fit $(N_5 + N_7)/N = cE^\beta$

$$c = B^{1/(1+\alpha)} \Delta^{(1-3\alpha)/(1+\alpha)} \left(\frac{2}{\pi\sigma}\right)^{(2\alpha)/(1+\alpha)}$$

and

$$\beta = (2\alpha)/(1+\alpha) \quad \Rightarrow \quad \alpha = \frac{\beta}{2-\beta} \tag{7.16}$$

With the values for β found in Sect. 7.8.1 for the combined 5- and 7-fold defect fractions this yields

	Experiment rI	Experiment rII
c	0.16	0.17
β	0.34	0.35
α	0.20	0.21

The calculated exponent α satisfies the condition $0<\alpha<1$. It was however a hypothesis that the system is fractal. Only if that is the case, (7.15) can describe its behavior correct. In 3D complex plasmas it was already found that a crystallization front could be characterized by a fractal dimension [21].

The other important assumption was the constancy of the surface tension σ in (7.14). For 3D systems there exist models for the temperature dependence of σ (e.g. [24, 25]). The Eötvös rule states a linear dependence $\sigma V^{2/3} = k (T_c - T)$ with a constant k, the critical temperature T_c and the molar volume V for a liquid pure substance. Ramsay and Shields improved this equation to $\sigma V^{2/3} = k (T_C - T - 6)$, taking into account a temperature offset of 6 K, which seemed to fit better to real data at low temperatures. Later Guggenheim and Katayama (and van der Waals) found $\sigma = \sigma_0 \left(1 - \frac{T}{T_C}\right)^n$ with a constant σ_0 and the empirical factor $n = 11/9$ which is valid for all organic liquids.

For 2D systems, no such dependencies are known yet. One could assume now that $\sigma = \sigma_0(T_C - T)$, similar to the Eötvös rule. Then the temperature dependence of the defects fractions in (7.15) would become $\frac{N_T}{N} \propto \left(\frac{T}{T_C-T}\right)^\beta$. If $T_C \ll T$ this would yield $\frac{T}{T_C-T} \to 1$ and N_T/N would not depend on T. For T_C within the measured temperature range, N_T/N would have a singularity and some feature like a kink should at least appear in the measured data. Both are not the case as one can see in the measured decay fractions. For $T_C \gg T$, the surface tension would be constant, and $\frac{T}{T_C-T} \to T/T_C$. The dependence $N_T/N \propto T^{\beta}$ follows as it was found earlier, therefore the assumption of a constant surface tension is consistent with the experimental findings.

7.9 Interpretation and Discussion

Now the nature of the transition is considered. That the transition exists is affirmed by the investigation of the structural properties (the point of melting in the following corresponds to the time when the pulse was initiated):

Translational Long-Range Order The pair correlation function $g(r)$ changes from a clearly liquid-like state with a correlation length ξ near 0 to a state with pronounced peaks and a $\xi \approx 10\Delta$ approximately 5 s after the melting was induced. The second peak splits at 1 eV. The correlation length covers a considerable range in the field of view, and the system can be considered to have long range translational order. $g(r)$ is the same at the end of the measurement as it was in the initial state.

Orientational Long-Range Order The initial state is characterized by a linear decay of $g_6(r)$. After melting, a short period with power law decay is found, which is soon replaced by an exponential decay. The correlation lengths ξ_6 increases from near to 0 to $\approx 6\Delta$ in a time span of 8 s after melting. Then it is replaced by the linear decay with a steeper slope as in the initial state. There is no long range orientational order in the regime of exponential decay. The linear decay was found to be caused by domain forming, and could be interpreted as some kind of intermediate range of orientational order, but clearly no long range order can be found.

Local Order The bond order parameter was high with $\Psi_6 > 0.8$ in average in the initial state, indicating a good hexagonal structure of the unit cells. It increases fast within 2 s from 0.34 to over 0.8 during recrystallization. The fraction of 5- and 7-fold defects is low (<5%) at the beginning and end of the measurement. It falls from over 25% within less than 2 s and reaches less than 5% approximately 4 s after melting, even before the translational order is restored. The defect condensation parameter indicates that most defects are organized in free dislocations in the initial state. After melting it falls from 3.5 (defect clustering) to 1.5 within 1 s and stays constant. From Fig. 7.15c, d one could see that aside from free dislocations, defects often have two defect neighbors, yielding chain-like structures. Qualitatively it was seen in Figs. 7.18 and 7.19 that also free dislocations are arranged in the chains, or grain boundaries.

All above findings clearly indicate that a transition occurs from a state of high disorder to an ordered state, characterized by long range translational order, short range orientational order and low defect fractions.

The experimental results are now compared with the possible theories:

KTHNY Theory A exponential decay of $g(r)$ indicates a liquid or hexatic phase. However, the strongly pronounced peaks and structure of $g(r)$ in the ordered state indicate the existence of long range order. ξ does further not decay exponentially with the energy, as it is predicted in the KTHNY theory. The orientational correlation function is never constant (solid state); when it decays with a power law (hexatic state) in the short period after melting, the exponent is >1 which is not compatible with $\eta_6 <$ 0.25. Elsewhere, $g_6(r)$ decays exponentially, meaning a liquid state according to KTHNY. The linear decay is not accounted for in the theory, because domain forming is not a feature there. The correlation length ξ_6 does also not depend exponentially on the energy (see the compilation in Sect. 6.1.2 for the temperature-dependence of ξ and ξ_6). The organization of defects in dislocations can be confirmed, but bound pairs of dislocations can not be identified.

According to the correlation analysis of the KTHNY theory, the system would be in a liquid state practically always. But the models proposed in that theory contradict the experimental findings, and it does not seem applicable to interpret the results in that context.

1st Order Transitions In general no indications for a first order transition can be found, since there are no discontinuities in any investigated parameter, or indications for latent heat. The density wave theory claims a discontinuity in the density. The interparticle distance, representing the density well, fluctuates short after melting, but this feature is attached to the shock of the electric pulse.

The grain boundary mechanism is interesting, since grain boundaries are found, but there appears no significant feature in the defect fractions in dependence on the kinetic energy which could indicate a 1st order transition. The exponential decay of $g_6(r)$—and absence of long range orientational order—was predicted in that theory to appear if there are no bound states between the grain boundaries. An investigation of bound states involves the summation of all Burgers vectors of the dislocations: If bound states exist, the total Burgers vector should be zero. This requires a state where no randomly distributed dislocations can cancel out by chance (no liquid state), and a record which captures all dislocations in the system, which was not the case here.

The modified Lindemann criterion of melting predicts a leap in the translational order parameter, which was not found, and it predicts that the orientational order persists at larger temperatures than the translational order, which is clearly not the case here. Note that the peak width obtained from the pair correlation function (shown in Fig. 7.10) is practically equal to $\gamma \Delta^2$ (γ is the Lindemann parameter). γ calculated from this relation for the presented data is always much smaller than the critical value 0.1 at the point of melting for 2D systems [26, 27]. Even at the highest energy it is still smaller by a factor of 10. That result would put the critical temperature far out of the measured range.

Kinetic Theory (Frenkel) The kinetic theory by Frenkel gives a direct dependence of the defect fraction on the system temperature. In the case of a system in thermo-dynamical equilibrium, an exponential Arrhenius law (6.7) is expected for the decay of the defect fraction with T.

Here the transition happens in a non-equilibrium regime, because the cooling is very rapid, and there is no time for thermodynamical equilibrium to become established. In comparison, in an experiment where the melting of a 2D plasma crystal was induced by different laser heating methods, stepwise to ensure an equilibrium

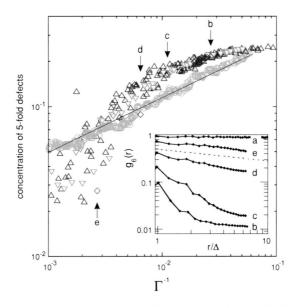

Fig. 7.24 Defect fractions vs. inverse temperature for several experiments with 2D plasma crystals, where melting was induced by laser heating, while the system was in thermodynamical equilibrium (*red diamonds*, *blue* and *green triangles*). The defect fractions decay according to an Arrhenius law (exponential). The *blue circles* are the defect fractions from the non-equilibrium recrystallization process presented earlier, and the *black line* shows the power law decay (figure from [4])

processes, an Arrhenius law with an exponential decay was found [4]. Figure 7.24 shows the exponential decay of defect fractions for several equilibrium situations (red diamonds and blue and green triangles), compared to the power law decay found in the experiments presented here (blue circles). The defect fractions are plotted vs. the inverse temperature.

The power law dependencies of several structural quantities (defect fractions, correlation lengths and bond order parameter) indicate a scale free transition from the disordered to the ordered state. Qualitatively, this behavior can be described by (7.15) based on the ratio of grain boundary length to domain size. This relation is valid only if the ratio has a fractal dimension, and if the surface tension σ is independent on the temperature. Assuming that, the power law exponent α can be calculated from the data, and compared with the theoretical prediction. Note that the theory only states an interval for α, not a fixed value.

Influence of the Grain Boundaries The reason for the scale free behavior is found in the mechanism of recrystallization. The defect condensation parameter showed that practically all 5- and 7-fold defects are organized in dislocations (Fig. 7.15), while the qualitative inspection of the particle system showed additionally that in the later stages most dislocations form chains which build up grain boundaries separating domains of different lattice orientation (Figs. 7.18 and 7.19).

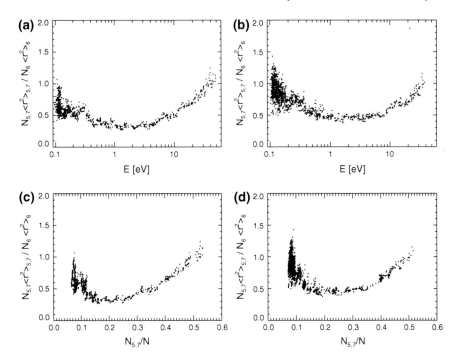

Fig. 7.25 Average MSD vs. kinetic energy E: **a** Experiment rI; **b** Experiment rII. Average MSD vs. total fraction of defects: **c** rI; **d** rII

It was shown in Fig. 7.21 that the average energy (represented qualitatively by the mean squared displacements) stored in a defect lattice site exceeds that stored in a hexagonal lattice site.

As a measure for the total energy of the grain boundaries in one frame, the total number of defects multiplied by their average mean square displacement, $\bar{r}_{5,7} = (N_5 + N_7)\langle r^2 \rangle_{5,7}$ with $\langle r^2 \rangle_{5,7}$ from (7.4), is calculated. The same is done for all 6-fold lattice sites (the bulk), yielding $\bar{r}_6$. Fig. 7.25 shows the ratio $\bar{r}_{5,7}/\bar{r}_6$ vs. the mean kinetic energy of the particles E (a, b), and vs. the defect fractions $(N_5 + N_7)/N$ (c, d). This ratio gives an impression of the total energy stored in grain boundaries compared to the total energy stored in the bulk.

The ratio starts at approximately 1 (corresponding to 50% of the energy is stored in defects and bulk, respectively) for high energies E and high defect fractions (right sides of the plots). It decreases with decreasing kinetic energy (decreasing defects fraction). At ≈ 1 eV, which corresponds to the time 5 s from the beginning of the measurement, and a defect fraction of ≈ 0.2 the behavior is reversed. The total energy stored in the boundaries increases again, however the number of particles constituting to them continues to fall: The system becomes cooler and more ordered into a hexagonal lattice, and the (fewer) defects eventually contain more thermal energy than the (much more numerous) particles in the well ordered domains.

As $E = 0.1$ eV is reached, the total energy stored in the boundaries, compared to that stored in all bulk particles, is only smaller by a factor of 0.6–0.8 in rI, and practically equal in rII, even though only 10% of the particles are organized in boundaries, i.e. approximately 40% (rI) and even 50% (rII) of the total system energy is stored in less than 10% of all the particles. At the turning point of the curves, the 20% defects already contain 25% (rI) and 30% (rII) of the total system energy.

The thermodynamical behavior of the system must therefore severely be determined by the small fraction of particles in the grain boundaries with the driving force being the high boundary energy. This illustrates the importance of defects.

The turning point of the curves, indicating the dominating influence of the boundaries, appears at a kinetic energy of approximately 1 eV. The behavior of other quantities at this point is investigated:

- For experiment rII: the exponent of the decay of the kinetic energy changes to a slower decay
- The second peak of the pair correlation splits up
- The fluctuation of the interparticle distance is at a turning point from increase to decrease
- The change of the slope of the peak width σ_0^2 vs. E appears at approximately 1 eV
- The bond correlation functions decays exponentially from 1 eV on
- At 1 eV: $\Psi_6 = 0.65$, $N_{5,7}/N = 0.08$
- The defect condensation parameter becomes constant with $S = 1.5$; the number of free dislocation exceeds the number of defects with two defect neighbors (crossing of blue and red lines in Fig. 7.15c, d)

The slower decay of the kinetic energy was found to be caused by local heating. The fluctuation of the interparticle distance was assigned to the remaining influence of the shock wave due to the mechanism of melting, and the slope change of the peak width was related to that feature. It is possible that until the energy of 1 eV the dynamical behavior of the particles is still affected by the propagating shock. From 1 eV on, where maybe domain sizes and boundary length have reached a critical threshold, the dynamics become more and more dominated by the grain boundaries, while the structural reorganization takes place.

The continuous increase of domain size, accompanied by the decrease of the boundary lengths (decrease of total number of defects) leads to a continuous change of the system state, and the scale free behavior ensues.

References

1. V. Nosenko, J. Goree, A. Piel, Laser method of heating monolayer dusty plasmas. Phys. Plasmas **13**, 032106 (2006)
2. Y. Feng, J. Goree, B. Liu, Solid superheating observed in two-dimensional strongly coupled dusty plasma. Phys. Rev. Lett. **100**, 205007 (2008)
3. M. Wolter, A. Melzer, Laser heating of particles in dusty plasmas. Phys. Rev. E **71**, 036414 (2005)

4. V. Nosenko, S.K. Zhdanov, A.V. Ivlev, C.A. Knapek, G.E. Morfill, 2D melting of plasma crystals: Equilibrium and nonequilibrium regimes. Phys. Rev. Lett. **103**(1), 015001 (2009)
5. M. Thomas H., E. Morfill G., Melting dynamics of a plasma crystal. Nature **379**, 806–809 (1996)
6. R.A. Quinn, J. Goree, Experimental test of two-dimensional melting through disclination unbinding. Phys. Rev. E **64**, 051404 (2001)
7. V.A. Schweigert, I.V. Schweigert, A. Melzer, A. Homann, A. Piel, Plasma crystal melting: A nonequilibrium phase transition. Phys. Rev. Lett. **80**(24), 5345–5348 (1998)
8. D. Samsonov, S.K. Zhdanov, R.A. Quinn, S.I. Popel, G.E. Morfill, Shock melting of a two-dimensional complex (dusty) plasma. Phys. Rev. Lett. **92**(25), 255004 (2004)
9. C.A. Knapek, D. Samsonov, S. Zhdanov, U. Konopka, G.E. Morfill, Recrystallization of a 2D plasma crystal. Phys. Rev. Lett. **98**, 015004 (2007)
10. B. Liu, J. Goree, V. Nosenko, L. Boufendi, Radiation pressure and gas drag forces on a melamine-formaldehyde microsphere in a dusty plasma. Phys. Plasma **10**(1), 9–20 (2002)
11. U. Konopka, Wechselwirkungen geladener Staubteilchen in Hochfrequenzplasmen. PhD thesis, Fakultät für Physik und Astronomie der Ruhr-Universität-Bochum (2000)
12. S. Nunomura, J. Goree, S. Hu, X. Wang, A. Bhattacharjee, Dispersion relations of longitudinal and transverse waves in two-dimensional screened Coulomb crystals. Phys. Rev. E **65**, 066402–111 (2002)
13. S. Nunomura, J. Goree, S. Hu, X. Wang, A. Bhattacharjee, K. Avinash, Phonon spectrum in a plasma crystal. Phys. Rev. Lett. **89**(3), 035001 (2002)
14. P.S. Epstein, On the resistance experienced by spheres in their motion through gases. Phys. Rev. **23**(6), 710–733 (1924)
15. L.V. Berezinskii, Destruction of long-range order in one-dimensional and two-dimensional systems having a continuous symmetry group I. Classical systems. Sov. Phys. JETP **32**(3), 493–500 (1971)
16. S.T. Chui, Grain-boundary theory of melting in two dimensions. Phys. Rev. Lett. **48**(14), 933–935 (1982)
17. V. Nosenko, S. Zhdanov, G. Morfill, Supersonic dislocations observed in a plasma crystal. Phys. Rev. Lett. **99**, 025002 (2007)
18. movie at http://www.mpe.mpg.de/~knapek/movies/index.html
19. D.R. Nelson, B. I. Halperin, Dislocation-mediated melting in two dimensions. Phys. Rev. B **19**, 2457 (1979)
20. V. Nosenko, S.K. Zhdanov, Dynamic of dislocations in a 2D plasma crystal. Contrib. Plasma Phys. **49**((4-5), 191–198 (2009)
21. M. Rubin-Zuzic,G.E. Morfill,A.V. Ivlev, R. Pompl, B.A. Klumov, W. Bunk, H.M. Thomas, H. Rothermel, O. Havnes, A. Fouquet, Kinetic development of crystallization fronts in complex plasmas. Nat. Phys. **2**, 181–185 (2006)
22. J. Frenkel, *Kinetic Theory of Liquids*. (Dover Publications, Inc, New York, 1955)
23. G.E. Morfill, A.V. Ivlev, Complex plasmas: An interdisciplinary research field. Rev. Mod. Phys. **81**, 1353–1404 (2008)
24. R. Eötvös, Ueber den Zusammenhang der Oberflächenspannung der Flüssigkeiten mit ihrem Molecularvolumen. Annalen der Physik und Chemie, **263**(3), 448–459 (1886)
25. E.A. Guggenheim, The principle of corresponding states. J. Chem. Phys. **13**((7), 253–261 (1945)
26. V.M. Bedanov, G. Gadiyak, On a modified Lindemann-like criterion for 2D melting. Phys. Lett. **109**((6), 289–291 (1985)
27. X.H. Zheng, J.C. Earnshaw, On the Lindemann criterion in 2D. Europhys. Lett. **41**((6), 635–640 (1998)

Chapter 8
Summary and Outlook

The work described in this thesis was concerned with the experimental investigation of the phase state of two-dimensional complex plasmas, consisting of micrometer sized plastic spheres immersed in an Argon plasma.

Error Analysis The data analysis was performed on recorded images containing the illuminated particles. Particle coordinates were extracted directly from this images by a tracking algorithm, which provides the possibility to observe physical processes on a fundamental kinetic level. The estimation of uncertainties due to the particle tracking procedure and due to pixel-noise in the recorded images was an important part of this work. Artificial images have been generated, containing particles as accumulations of "illuminated" pixels, and a random particle motion has been simulated in series of consecutive images. The influence of different quantities of pixel-noise and particle–image sizes (corresponding to the number of illuminated pixels per particle) was investigated. The procedure yielded estimates for the uncertainties which could be transferred to the experimental data, and enabled to identify the limits of resolvable distances. This allowed the validation especially of the measurement of particle velocities and displacements above the error threshold. The uncertainties were considered throughout the data analysis for all experiments individually. Only by considering this errors, the scientific interpretation of the acquired data gains a physical meaning.

Two kinds of experiments have been performed, addressing different aspects of the state of the two-dimensional particle system.

Estimation of the Coupling Parameter In the first experiments, the dimensionless coupling parameter Γ, defined as the ratio of mean potential to mean kinetic energy, was measured in a two-dimensional plasma crystal recorded at a high spatial and temporal resolution. Γ characterizes the phase state of the system, and can usually be calculated from the quantities particle charge and temperature, and the screening parameter κ which defines the modification of the interparticle potential due to charge screening. Here a new method was presented to obtain Γ from the directly accessible particle coordinates by using the relations between the caged particle motion and the potential and kinetic energy of the particles. It was shown that the results of the

C. A. Knapek, *Phase Transitions in Two-Dimensional Complex Plasmas*,
Springer Theses, DOI: 10.1007/978-3-642-19671-3_8,
© Springer-Verlag Berlin Heidelberg 2011

measurement are in good agreement with the results of the conventional method for the calculation of Γ from particle charge and temperature. This new method therefore provides an alternative way for the estimation of the coupling parameter without the need of additional experimental effort for charge and temperature measurement.

Recrystallization The second experiments were dedicated to the phase transition of a two-dimensional complex plasma. A single-layer plasma crystal was melted by a short electric pulse, and its undisturbed recrystallization was studied time-dependent with regard to the kinetic energy of the particles, and several structural properties. Due to the high temporal resolution of the recording, the mechanism of the phase transition could be studied on a fundamental kinetic level, and at the same time allows for the first time a connection between the thermodynamical state and the particle motion itself. It was found that a disorder–order transition takes place (confirmed by the development of structural properties), from a state without long-range order to a state characterized by long-range translational order, but only short range orientational order. The transition happens in a non-equilibrium regime, and power–law dependencies of the defect fractions, the bond order parameter (as a measure of the local crystal structure) and the translational and orientational correlation lengths (measuring the range of the respective order) on the kinetic particle energy were found during the crystallization process. This indicates a scale-free transition, in comparison to the exponential dependence (Arrhenius law) expected in a transition of a system in thermodynamical equilibrium.

A qualitative analysis showed that during the stages of the recrystallization small domains of uniform, hexagonal lattice orientation formed. The domains were separated by grain boundaries consisting of strings of dislocations, and the lattice orientation often changed abruptly across such boundaries, thus destroying the long-range orientational order. During the rapid cooling, the domains continuously grew in size, and the number of defects (and the boundary length) decreased. It was further shown that a considerable amount of the total thermal energy of the system was located in the grain boundaries: up to 50% of the energy was stored in less than 10% of the particles. Therefore the thermodynamical behavior of the cooling plasma crystal has to be strongly determined by the high energetic grain boundaries made from dislocations, leading to the observed scale-free transition.

The mechanism of the transition could not be explained by conventional theories of two-dimensional melting, such as dislocation-mediated melting or theories of first-order transitions. A theory by Frenkel [1], based on the relation between the domain size and the length of its boundary, explains the observed power–law dependence of the defect fractions under the assumptions that the grain boundaries have a fractal dimension, and that the surface tension at the domain border does not depend on the kinetic energy. From the present data the dimension of the boundaries could not be concluded, because the observed system is too small and boundaries vanish out of the field of view. The assumption of a constant surface tension was consistent with the experimental data, but yet there exist no theories on this subject for two-dimensional systems.

Outlook In future experiments, the results of the investigation of measurement uncertainties can be used to adjust the experimental parameters, especially of the data acquisition (camera settings and particle illumination), so that the range for particle energy measurements can be extended to lower values.

Another open question is the dimensionality of the grain boundaries, and the possible temperature dependence of the surface tension in two-dimensional systems. Answering that questions would allow a substantiated conclusion as to the validity of the kinetic theory for the two-dimensional complex plasma.

Reference

1. J. Frenkel, *Kinetic Theory of Liquids* (Dover Publications, New York, 1955)

Chapter 9
Appendix A: Estimation of Uncertainties in the Particle Coordinates

The tracking algorithm uses an intensity weighting center-of-mass method and was introduced in Sect. 3.2. The basic equation for the algorithm is given here as a short reminder. A particle position (x, y) in an image is calculated as

$$x = \frac{\sum_{i=1}^{n_x} x_i I_i}{\sum_{i=1}^{n_x} I_i}, \quad y = \frac{\sum_{i=1}^{n_y} y_i I_i}{\sum_{i=1}^{n_y} I_i} \tag{9.1}$$

The sums run over the number of pixels identified as a particle in x- and y-direction, n_x and n_y, respectively. I_i are the intensity values of pixel i.

The tracing of particles in consecutive frames yield the time dependent particle trajectories. A particle k in a frame $t + dt$ is identified to be particle i from frame t, if the distance between both coordinates $\mathbf{x}_k(t + dt) - \mathbf{x}_i(t)$ is smaller than a user supplied value. This value has to be chosen for a data set according to the expected particle displacements: it has to be larger than the expected velocity, but much smaller than the average distance between particles to ensure that the particle assignation is correct. After assigning all particles, the displacements, or particle velocities, for particle k are

$$\mathbf{v}_k(t) = (\mathbf{x}_k(t + dt) - \mathbf{x}_k(t))/dt \tag{9.2}$$

dt is the time step in seconds between two images, or in case of artificial data sets, it is equal 1.

The next Sect. 9.1 treats the uncertainties arising from the algorithm itself by tracking a single artificial particle at different subpixel positions and comparing its real position with the tracked position. Section 9.2 extends the analysis to a larger space of parameters as to particle-image size, pixel-noise and further investigates the influence of the error on quantities depending on spatial particle displacements. A statistical evaluation of the errors is performed. Section 9.3 addresses the problem of extracting the quantity of the pixel noise from real data.

C. A. Knapek, *Phase Transitions in Two-Dimensional Complex Plasmas*,
Springer Theses, DOI: 10.1007/978-3-642-19671-3_9,
© Springer-Verlag Berlin Heidelberg 2011

9.1 Single Artificial Particle

The error of the tracking algorithm can be analyzed by tracking an artificial particle with a defined central position and comparing the outcome of the tracking with the original value. The particle was designed as a two-dimensional Gaussian intensity distribution $I(x, y)$:

$$I(x, y) = 255 \times \exp\left[-\frac{(x - x_c)^2 + (y - y_c)^2}{2\sigma^2}\right] \tag{9.3}$$

$I(x, y)$ was calculated for integer positions (x, y) to simulate the pixel grid, with $\{x_c, y_c\} \in \mathbb{R}$ being the real center of the distribution and σ the Gaussian half-width. The maximum peak height of 255 corresponds to 256 greyscales (0 counts as one value). The matrix containing $I(x, y)$ was then written into an image with 8 bit color depth. In this process the intensity values are scaled to the color space of 256 greyscales and rounded to integers. This simulates images comparable in their properties to those recorded by the camera in the experiments. The left panel of Fig. 9.1 shows an example artificial particle with $\sigma = 1$ and $\{x_c, y_c\} = \{10.71, 10.71\}$.

100 images with the artificial particle were now generated, in which the center was shifted consecutively from frame to frame, starting from the center of pixel $(0, 0)$ in direction of the center of pixel $(1, 1)$ in steps of 0.01 pixels per coordinate and frame. To simulate particles of different size, this was done for each value of σ from 0.2 to 1 in steps of 0.1, i.e. nine sets of 100 images per set were generated. The number of "illuminated" pixels per particle for different σ is shown in the right panel of Fig. 9.1. Note that while the particle moves through the pixel, the number of illuminated pixels can change within one set of images for one value σ, because an adjacent pixel can get a non-zero intensity value as the particle moves toward it, while all formerly illuminated pixels still have a non-zero intensity.

The particle of each set was then tracked with the intensity weighted center-of-mass method described above. Figures 9.2 and 9.3 show the results for all σ. In the first column the tracked position is plotted vs. the real position, which in this case is the continuous sub-pixel position. The second column shows the deviation between tracked and real particle position vs. the real position. The number of pixels of the artificial particle in dependence on its real position can be seen in the third column. One can see clearly that the number of illuminated pixels changes for each σ, depending on the real particle position within the pixel. Also note that though the same number of pixels might appear for different σ, the errors differ in that cases. The error depends predominantly on the parameters of the intensity distribution, not on the actual number of pixels per particle. On the other hand, the geometrical alignment of pixels, i.e. how many pixels are illuminated in x- and y-direction, is not considered here. But still the number of pixels can be a hint on the quantity of the error, especially for very large or very small particle images.

Next it stands out that for $\sigma > 0.6$ the absolute deviations are rather small <0.01 pixel, and do not depend on the sub-pixel position. For $\sigma = 0.6$ one can see a slight dependence, but the error is still small. The dependence grows rapidly for smaller

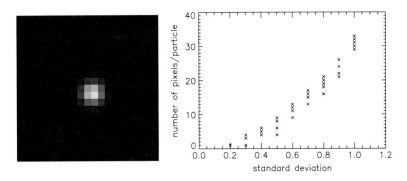

Fig. 9.1 *Left* Example of an artificial particle with a Gaussian intensity profile centered around $\{x_c, y_c\} = \{10.71, 10.71\}$ with a width of $\sigma = 1$. *Right* Number of pixels per particle vs. standard deviation of the Gaussian intensity distribution used to create the particles

values $\sigma = 0.5 - 0.2$, while simultaneously the maximal value of the error rises above 0.01 pixel. For $\sigma = 0.2$ the single particle has only one illuminated pixel, which leads to the so-called pixel-locking effect: the total intensity of the particle is either located in the one pixel or in the next, though the particle in fact moves through the pixel. This causes the tracked position to be fixed in one pixel center until the particle crosses the border between two pixels. Then a jump of $+1$ pixel appears in the tracked coordinate. Pixel-locking can be significant also for larger particles, as will be shown later.

In summary, for particles with more than ≈ 10–15 illuminated pixels the error of the tracking algorithm is small enough to be neglected, and besides that it has characteristics of a statistical error, and its magnitude is independent of the number of pixels. For smaller particle images, the error is systematic and depends strongly on the real position within the pixel. Fig. 9.4 shows the maximal error one has to consider (red dots). It is the maximum absolute value of the deviation of tracked to real particle position in the middle columns of Figs. 9.2 and 9.3. Since in general nothing is known about a particles real position, this worst case is always possible and is the accuracy with which the particle positions can be estimated. The blue dots in Fig. 9.4 are the mean of the absolute value of all possible errors within a pixel for a particle of a certain size, with the corresponding standard deviation shown as error bars. Since the errors are not statistically distributed, the calculation of the mean is not a good representation, but it still gives an idea of the spreading of the possible errors. Finally, the black solid line is $1/\sqrt{np}$. This is the error $\delta\bar{x}$ one would expect for a particle consisting of np pixels, when the position $\bar{x}$ is calculated as the mean of all pixels regardless the intensities, and the accuracy of each pixel is $\delta x_i = 1$ px:

$$\delta\bar{x} = \sqrt{\sum_{i=1}^{np}\left(\frac{1}{np}\delta x_i\right)^2} = \frac{1}{np}\sqrt{\sum_{i=1}^{np}1} = \frac{np}{\sqrt{np}} = \frac{1}{\sqrt{np}} \qquad (9.4)$$

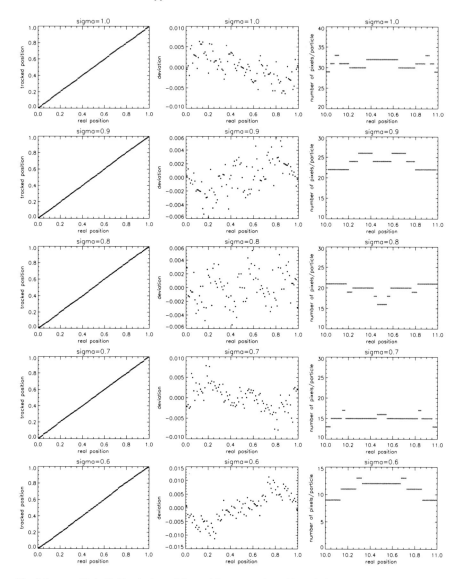

Fig. 9.2 $\sigma \in [0.6, 1]$: Tracked particle position vs. real position (*left column*); deviation of the tracked particle position to the real position vs. the particles subpixel position (*middle column*); number of illuminated pixels vs. particle subpixel position (*right column*)

The error in dependence on the number of pixels is high for small particles but falls rapidly as demonstrated in Fig. 9.4 for both maximum and mean error. The highest values for particles with only a few pixels origin from the strong influence of pixel-locking, which can only be avoided by providing measurements with large enough n_P.

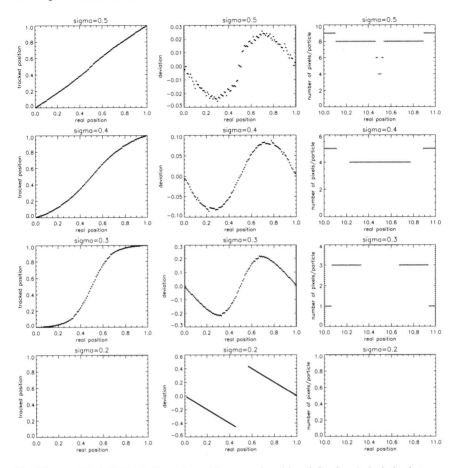

Fig. 9.3 $\sigma \in [0.2–0.5]$: Tracked particle position vs. real position (*left column*), deviation between both vs. subpixel position (*middle column*), number of illuminated pixels vs. subpixel position (*right column*). For $\sigma = 0.2$ (*bottom line*) the particle is located at $y = 0$ ($x = 0$–0.4) and $y = 1$ ($x = 0.6 – 1$), and $n_P = 1$ always (for x = 0.4–0.6 the particle was not tracked)

Since the error of the tracking algorithm is a systematic error, its propagation into quantities calculated from the particle positions can not be obtained by Gaussian error propagation. Of special interest here are particle velocities or in general the calculation of a distance between two positions, i.e. a spatial displacement. In that case the systematic errors are subtracted from each other as the following example shows for the calculation of a distance Δx between two coordinates x_1, x_2 with the respective errors δx_1, δx_2:

$$\Delta x + \delta(\Delta x) = (x_2 + \delta x_2) - (x_1 + \delta x_1) = (x_2 - x_1) + (\delta x_2 - \delta x_1) \quad (9.5)$$

The (signed) systematic errors in the coordinates could either cancel out or combine to the maximal possible value for the error of Δx, depending on the locations of x_1 and x_2 within a pixel. This kind of error propagation especially applies for the

Fig. 9.4 Maximum error
(*red dots*) and mean error
(*blue dots*) vs. number of
pixels per particle n_P. The
error bars for the mean error
are the standard deviation of
the values used for the
calculation of the mean. The
black line is $1/\sqrt{n_P}$

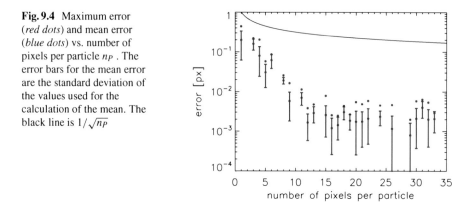

calculation of velocities v as the spatial displacement of a particle between the times t and $t + dt$, scaled with $1/dt$. The magnitude of the error of the velocities is expected to be dependent on the magnitude of the velocity itself, $\delta v \propto v$: With higher velocities, the particles move a larger distance from frame to frame and it is more likely that they are not close their original sub-pixel position. Since the error in the coordinates is systematic, a small motion within one pixel does not produce a large error (errors are subtracted in the calculation of δv), because the errors are nearly the same. For larger velocities, the probability increases that two large errors with maybe even opposite sign are subtracted, which leads to high errors in the velocities. This will be investigated later in Sect. 9.2.

The next section will address the behavior and the influence of the error in the case of statistical analyses of the dynamics of larger numbers of particles. Further, the impact of pixel-noise imposed on the images, which is inevitable in a real record due to the finite temperature of the CCD or CMOS chip in the camera, will be investigated.

9.2 Artificial Particles at Random Positions

To obtain a better statistic, sets of images with 2500 artificial particles per image were generated in the following manner:

At first the particle center positions were placed on the nodes of a square grid with the edge length 20 px, $\mathbf{x}_{i,\text{fixed}} = 20\binom{k}{n}$ with $k, n \in [1, 50]$ and i being the particle number. To these coordinates, random numbers $\mathbf{x}_{i,\text{random}}$ out of the interval [0.00, 1.00] with two decimal places were added. The random numbers were drawn from an uniform distribution. This method ensures that a lot of different subpixel positions $\bar{\mathbf{x}}_{i,0}$ are occupied and the uniqueness of the error can be tested sufficiently.

$$\bar{\mathbf{x}}_{i,0} = \mathbf{x}_{i,\text{fixed}} + \mathbf{x}_{i,\text{random}} \tag{9.6}$$

Above equation defines the particle positions in the first image.

A set of ten consecutive coordinate arrays was then generated by successively adding a displacement ξ to the coordinates $\bar{\mathbf{x}}_{i,0}$. The displacement was taken at random from a Gaussian distribution with the width $\Delta\xi$:

$$f(\xi) = \frac{1}{\sqrt{2\pi}(\Delta\xi)} \exp\left(-\frac{\xi^2}{2(\Delta\xi)^2}\right) \tag{9.7}$$

The width is given in units of pixels. It can be compared with a particle temperature $k_B T = (\Delta\xi(dr)/(dt))^2 m$ with the particle mass m and the spatial and temporal resolution dr and dt as scaling factors. The particle center coordinates $\bar{\mathbf{x}}_i(t)$ in an image at time t are now:

$$\bar{\mathbf{x}}_i(t) = \bar{\mathbf{x}}_{i,0} + \sum_{j=1}^{t} \xi_j \tag{9.8}$$

The summation was performed for $t \in [0, 9]$, yielding 10 consecutive coordinate arrays with coordinates for 2500 particles each.

The final coordinates $\bar{\mathbf{x}}_i(t)$ are used in (9.3) to generate the image matrices containing all particles, centered at the respective $\bar{\mathbf{x}}_i(t)$, and with the shape of a Gaussian intensity distribution for given values of the distribution width (or particle size) σ.

To take into account different particles sizes and displacements, for each $\sigma \in [0.2, 1]$ in steps of 0.05 px (17 different σ) and for 12 widths $\Delta\xi$ of the displacement distribution from (9.7) out of the interval $[0.014, 0.09]$ px, a set of ten images was generated, giving a total of 204 data sets.

The range of σ covers particle sizes of 1 pixels up to over 30 pixels. The width $\Delta\xi$ was chosen such that after scaling the outcomes with the particle mass and spatial and temporal resolution used in the experiments, the displacement becomes comparable to the range of displacements (or velocities) usually observed in the experiment, i.e. for a spatial resolution of 0.034×10^{-3} m/px, a mass of 6.14×10^{-13} kg and a time step between frames of 0.006 s, the widths correspond—in units of temperature—to a range 0.024–0.99 eV.

Error in the Coordinates All set of images were tracked by the particle tracking routine described in the beginning of the appendix, and the difference between tracked and real coordinates was analyzed. The real coordinates are the particle centers which where calculated for each frame and each particle as explained above and stored in an array. The normalized histograms (divided by bin size and total number of counts) of the error in the particle coordinates are shown in Fig. 9.5 for the different σ. From $\sigma = 0.65$ on, the errors have a Gaussian distribution centered around zero.

The impact of pixel-locking becomes visible in xy-maps of the subpixel parts of all tracked coordinates. This is shown in Fig. 9.6 for each value of σ. Pixel-locking is evident for $\sigma = 0.2$ and $\sigma = 0.25$. For $\sigma = 0.3$–0.35 there still appear some preferred structures, while for $\sigma \geq 0.4$ the distribution of subpixel positions quickly becomes uniform.

Error in the Displacements Displacements are calculated as the subtraction of the particle positions from two consecutive frames. The real displacements from frame

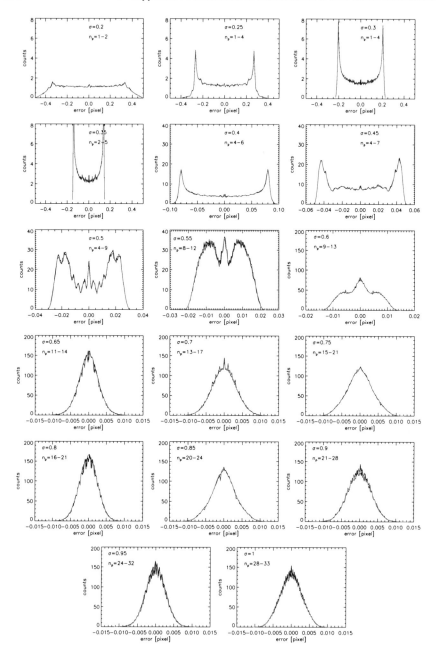

Fig. 9.5 Histograms of errors in the particle coordinates *x* (*black*) and *y* (*red*) for different widths σ of the particle intensity distribution. The ranges of the number of illuminated pixels per particle are shown inside the figures

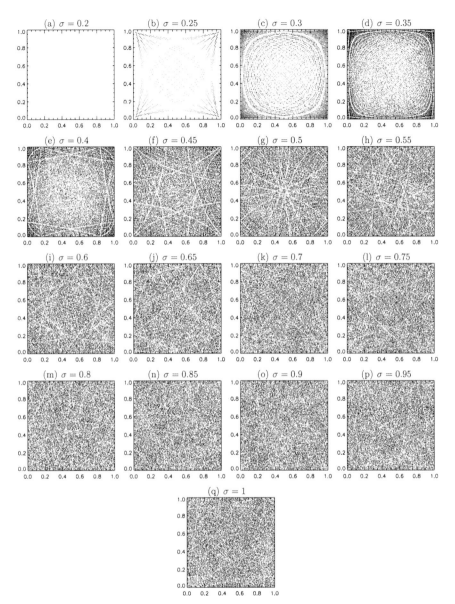

Fig. 9.6 Maps x vs. y of the subpixel parts of the tracked coordinates of artificial particles for different width σ of the particle intensity distribution. For $\sigma = 0.2$ (**a**), all points are located on the x- and y-axes

to frame are known, they can be obtained from the subtraction of the respective real positions. The tracking and tracing routines for experimental data were used on all 204 sets of artificial images, and yield the measured particle displacements.

Then the error of the displacement is calculated as the difference between the measured and the real displacements for each particle and each time step. This gives 22500 values for the error per data set, which will be analyzed statistically.

The real displacements are Gaussian distributed by definition, because they were calculated by adding values drawn from a Gaussian distribution to an initial position. The histograms of real displacements were calculated anyway and fitted by Gaussians, with the goodness of fit $\chi^2 \approx 1$, as it is expected.

An example for the histograms of measured displacements is shown in Fig. 9.7. For small particles ($\sigma < 0.35$) the histograms look Gaussian distributed, but with one additional, high central peak (one bin in the histogram) at the position displacement ≈ 0. If a Gaussian distribution is fitted including the central peak, the obtained width of the distribution will be very small. Such a peak can usually be omitted in a fit, because it just shows that most values were practically zero and are of no interest. Here the peak is clearly the effect of pixel locking and the resulting, apparent lack of particle motion from frame to frame. This feature of the central high peak depends on the particle size only, not on the quantity of the particle motion (i.e. the quantity of the chosen particle "temperature" which simulates the motion). The measured displacement histograms were fitted by Gaussians, excluding the central peak.

Now the error of particle displacements will be investigated. Figure 9.8 shows example histograms of the error for all values of σ for the data sets with the lowest $\Delta\xi$, i.e. the data were particles move the slowest. The histograms were obtained for the components v_x (black) and v_y (red) of the displacement separately. The value of σ and the range of the number of pixels per particle, n_p, is given in each plot window. The error histograms can be fitted well by Gaussian distributions, with a χ^2 between 1–25, where the higher χ^2 appear for the smaller particles. This fit was performed on all 204 data sets, yielding 14×17 fit parameters (width and mean of the fitted Gaussian distribution) for the error histograms, and the same for the fit to the histograms of real displacements and measured displacements. The widths of the latter two distributions, interpreted as velocity distributions, would determine a particle temperature, while the width of the error distribution has to be interpreted as a kind of white noise.

The scaling of the error with $\Delta\xi$ was already addressed in the former chapter and manifests itself in broader error distributions with increasing displacement $\Delta\xi$ from frame to frame, but they stay Gaussian with approximately the same goodness of the fit.

Figure 9.9 shows the widths obtained from the Gaussian fits for real and measured displacements, and for the errors in comparison. The values for one σ are plotted versus the width $\Delta\xi$ of the distribution which defined the particle motion. The scaling of the error with $\Delta\xi$ was already addressed in the former chapter and manifests itself in broader, but always Gaussian, error distributions with increasing $\Delta\xi$. The goodness of the fit χ^2 stayed approximately the same as given above.

The width of real displacements (black diamonds) is always equal to $\Delta\xi$. Until $\sigma = 0.3$, the width of measured displacements (blue crosses) is larger than the real value. The width of the error distribution (red dots) is very large and practically equal to $\Delta\xi$. This range of σ correspond to the range of particle sizes until which the

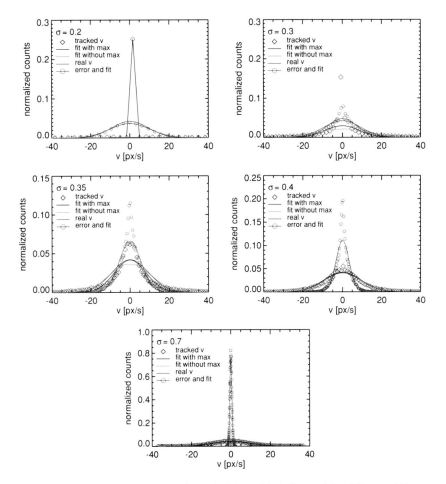

Fig. 9.7 Examples of histograms of particle velocities v (*black diamond*) for different widths $\sigma = 0.2, 0.3, 0.35, 0.4, 0.7$ of the particle intensity distribution. The *black* and *green lines* are fits with and without considering the maximum peak, respectively. The *red circle* and line is the histogram of the error, the *blue line* shows the histogram of real velocities

high central peak appears in the histogram of displacements. Since the peak was not fitted, the width of measured displacements is much higher than the real width. It was tested what happens if the peak is included in the fit: the measured width becomes much lower than the real one and with increasing σ approaches the real width from below, and exceeds the error considerably not until $\sigma = 0.4$. One way or the other, it is never the real width of the displacement distributions, because the error is of the magnitude of the quantity which should be measured. Only from $\sigma = 0.35$ on, the expected width is obtained (the black and blue line become equal, and the quantity of the error drops). This corresponds to a particle size larger than $\approx$2–5 pixels, where 2 pixels in an exception. In Fig. 9.10 the percentage of the occurrence of a certain

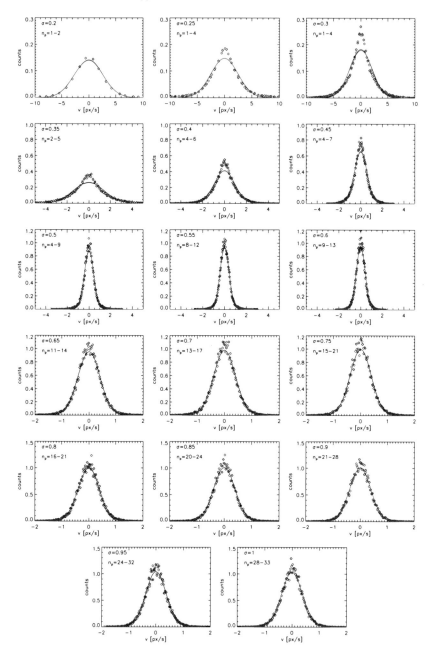

Fig. 9.8 Histograms of errors in the particle velocities v_x (*black*) and v_y (*red*) for different widths σ of the particle intensity distribution. The ranges of the number of illuminated pixels per particle are shown inside the figures

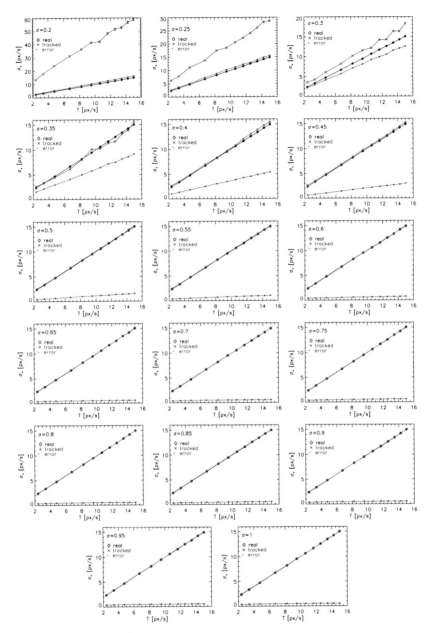

Fig. 9.9 Standard deviations of Gaussian fits without considering the maximum peak to the particle velocity distributions for real (*black diamonds*) and tracked (*blue crosses*) velocities and for the distribution of the deviation of tracked from real velocities (*red dots*). Error bars (1-σ uncertainty of the fit) are of the size of the plot symbol

Fig. 9.10 Percentage of
occurrence of a certain
number of illuminated pixels
for particle sizes defined by
the width $\sigma = 0.35$ of the
Gaussian intensity
distribution

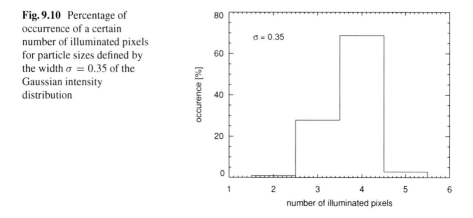

number of illuminated pixels for $\sigma = 0.35$ is shown. Over 60% are at 4 pixels, and
the particle consists of 3–4 pixels most of the time.

In conclusion, though the error caused by the tracking algorithm is systematic for
the particle coordinates and can take on values of up to 0.5 pixels in the worst case,
the error which follows for particle displacements or velocities can be described as
a statistical error. If particles are too small, pixel locking is still a problem. Further,
if the particle size is >3 pixels, the error in the displacements seems not to have
any influence on the measured width of the displacement or velocity distribution.
Therefore it should be possible to e.g. calculate particle temperatures or kinetic
energies from the width of such distributions, even if the expected error in the particle
coordinates is large.

9.3 Pixel-Noise

The uncertainty caused by the tracking itself is not the only source of errors. Images
taken with a digital camera are subject to noise which originates from the finite
temperature of the chip. This temperature heats the pixels in the chip and causes small
currents to occur which in turn simulate an incoming intensity. Since the source of this
intensity is Gaussian (white noise), the pixel-noise appears as Gaussian distributed
intensity values superimposed on all pixels, which are characterized by a certain
width of their distribution. Due to the random nature of this noise, it is not possible
to substract it. The only possibility is to learn about the quantity of the noise and the
restrictions it imposes on resolving particle positions and particle motion.

Noise is simulated in the artificial images from the previous section by adding an
array of random intensity values to the original image arrays. Then each pixel has
the intensity $\tilde{I}(x, y) = I(x, y) + I_{noise}$. Noise levels of 2–20 in steps of 2 in units
of intensity (with a maximum intensity of 255) where chosen, yielding ten noise
levels. Each data set from the former section with 12 different values of particle
displacement and 17 different values of particle size, was superimposed with each
of the ten noise levels. Additionally, data sets for all 17 particle sizes for two more

simulated particle motions 0.202 px and 0.285 px (corresponding to 5 and 10 eV) where generated and analyzed as noiseless and noisy data. This gives a total of 238 data sets without noise, and 2380 with different noise levels, each set consisting of ten images. For each data set, the original, real particle coordinates are known. Then the same analysis as in the last chapter was performed on all 2380 data sets.

Error in the Coordinates Particle coordinates were tracked for each set of images and the deviation of tracked positions to real position is calculated. Instead of analyzing the data in dependence on the width σ as it was done before, the number of pixels per particle is taken as the particle size. In real data, the width of the intensity distribution of the particle image is not known, the only quantity which can be estimated easily is the number of illuminated pixels per particle. Now the error in the coordinates is presented as the histograms of the error for particle sizes of 2–11 pixels for all noise levels, and for the data without noise, in Fig. 9.11. Noise levels are distinguished by colors. The noiseless data (dark violet lines) have clearly a non Gaussian distribution, as already found in former chapters. For small particles and low noise levels, the histograms are also not Gaussian. The shape of the distributions becomes more and more normal with increasing noise and particle size. Particle with sizes $N_P \geq 5$ px and noise levels of ≥ 6 have a Gaussian shaped histogram of the errors, while from $N_P \geq 10$ px on, any noise level leads to the statistical error distribution. An exception exists for the particle size $N_P = 4$: a noise level of 4 already causes the error histogram to have a Gaussian shape. This is likely due to the geometrical influence a central symmetric quadratic shape of 4 pixels has on the particle center calculation.

To obtain a rough estimate on the quantity of the errors, Gaussian distributions were fitted to all histograms, and their standard deviations are plotted in Fig. 9.12, representing the mean error. The non-normal shape of the histograms at low noise levels and small N_P has to be kept in mind when interpreting or using this error values.

Error in the Displacements The errors for particle displacements are calculated as before by tracing particles from frame to frame and subtracting the measured displacements from the real displacements. The histograms of the errors are analyzed. An example for the shape of the histograms is shown in Fig. 9.13 for mean particle displacements from frame to frame of $\Delta \xi = 0.029$ px. All histograms could be fitted well by Gaussian distributions. The distributions for noiseless data were already investigated in the last chapter. The noisy data yield a statistical error, increasing with increasing noise level and particle size. For other values of the defined particle motion $\Delta \xi$, the histograms were always Gaussian, too, and fits yield the widths of the error histograms as the estimate for the average error. The error in particle displacements is statistical, as it was already found for the noise-free data.

The widths of all error distributions are plotted in Figs. 9.14 and 9.15 vs. the noise levels, color-coded for the different particle sizes. One plot panel represents one particle motion range in units of px. For slow particles, i.e. particle whose coordinates were calculated with a low $\Delta \xi$, the error increases with increasing noise level. For faster particles, the error at low noise levels seems to increase especially for small particles, until at the highest displacements from frame to frame

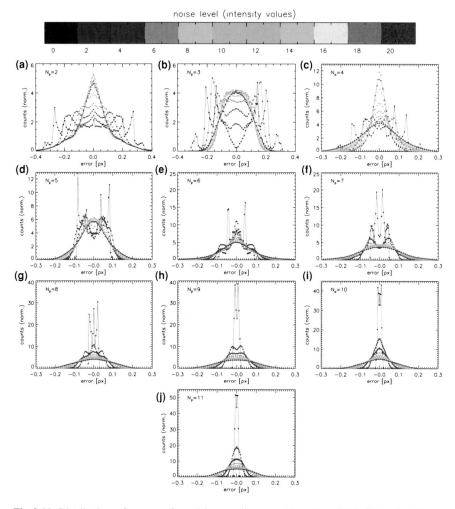

Fig. 9.11 Distribution of errors of particle coordinates with noise added. Noise levels are 0, 2, 4, 6, 8, 10, 12, 14, 16, 18, 20 (in units of intensity with a maximum intensity of 255)

(corresponding to 5 and 10 eV in the example in Sect. 9.2), particle with sizes ≤4 pixels have the largest errors in displacements at noise levels <6. Particles with only 2 pixels (dark violet line) stand out with a very small error for small displacements. This is again the effect of pixel-locking, leading to a lot of measured displacements close to 0. For low noise levels, the error is therefore practically of the magnitude of the real displacements, while for large noise levels, the motion is completely masked by the statistical, superimposed noise, and the errors seem not to change very much with increasing noise.

To emphasize the effect of the particle displacement, which in real data would be the particle temperature, the same error values are plotted vs. the average particle

Fig. 9.12 Mean error in the particle coordinates vs. pixel-noise level for different particle sizes. Particle sizes are defined as the number of illuminated pixels per particle. The mean error is estimated from the Gaussian widths of the histograms

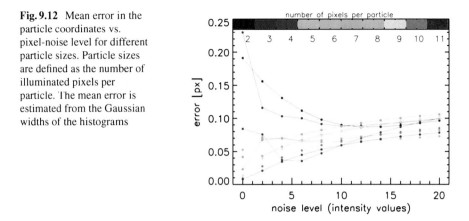

displacement separately for each noise level in Figs. 9.16 and 9.17. The errors for different particle sizes are again color-coded.

One can see from all four figures in summary, that:

1. The error in general increases with decreasing particle size (see different colored curves)
2. The error increases with increasing noise level, with a steeper increase at lower noise levels
3. The error for small particles is influenced mostly by the particle "temperature", and of the magnitude of the real particle displacement from frame to frame as an effect of pixel locking. This leads to the very high error for low noise and small particles with a high temperature (best seen in Fig. 9.15f)
4. At large noise levels, the errors increase very slowly with further increasing noise

The most important result of the simulations and statistics is that the errors of particle coordinates and displacements, or velocities, become statistical in the presence of pixel noise.

9.3.1 Noise in Real Images

To use the results of the former chapters for the magnitude of uncertainties, one needs to know the particle size in pixels and the value of the pixel noise. Then the error for a real measurement is given by the error calculated above for the respective values. The number of illuminated pixels per particles is determined during particle tracking. Two methods to estimate the quantity of the pixel noise in real data are introduced in the following. The first is the simple case applicable for images with a background noise. The second method is needed in case the background was cut off during recording.

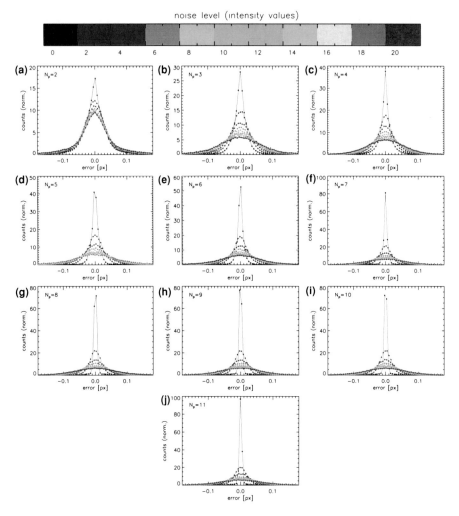

Fig. 9.13 Distribution of errors of particle displacements from frame to frame with noise added. Noise levels are 0, 2, 4, 6, 8, 10, 12, 14, 16, 18, 20 (in units of intensity with a maximum intensity of 255). The average particle displacement was $\Delta \xi = 0.029$ px

9.3.2 Pixel-Noise from Background

If there is background noise in the images in regions between illuminated particles, the noise can be estimated as the width of the distribution of intensity values below the threshold chosen for particle identification. This is done for a larger amount of consecutive images of a measurement to get good statistical results. The widths obtained from the individual frames are averaged afterwards.

A method to decrease pixel noise is to average consecutive images as to their intensity values. Each pixel is subject to the same quantity of pixel noise, represented

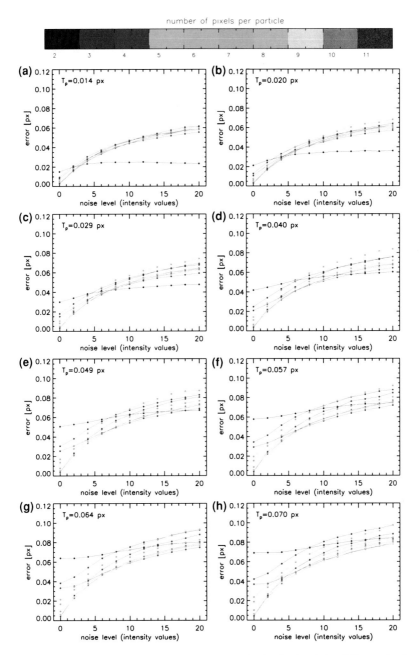

Fig. 9.14 Average error of particle displacements vs. pixel-noise level for different simulated particle motion of the quantity 0.014 px to 0.070 px. Number of pixels per particle are color-coded

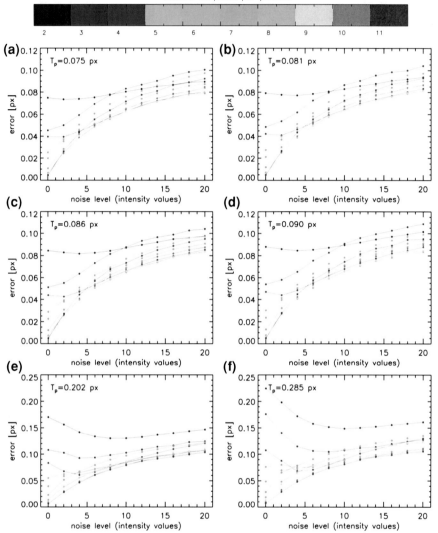

Fig. 9.15 Average error of particle displacements vs. pixel-noise level for different simulated particle motion of the quantity 0.075–0.285 px. Number of pixels per particle are color-coded

as random values taken from the same Gaussian distribution. The addition of the intensity values of two pixels is equivalent to the convolution of two Gaussians with the same width. The Gaussian distribution is invariant against convolutions, therefore this addition leads to a new, smeared out Gaussian noise distribution with a width scaled by a factor of $\sqrt{2}$. By not only adding images, but averaging them, all intensity values, and also the width are divided by the number n of images used for averaging,

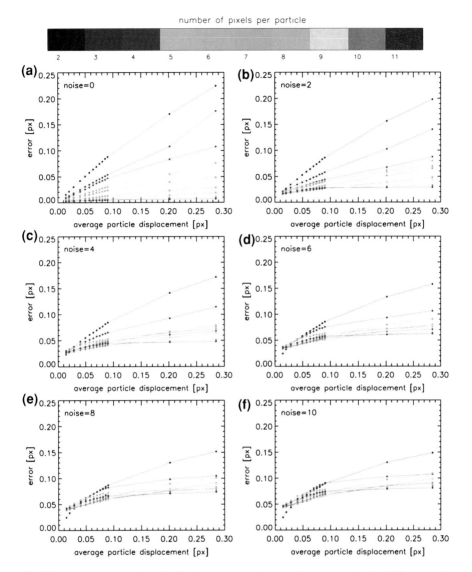

Fig. 9.16 Average error of particle displacements vs. simulated particle motion for different pixel-noise levels 0–10 in units of intensity. Number of pixels per particle are color-coded

and the resulting noise distribution has the original width divided by $\sqrt{n}$. Therefore, averaging consecutive images can significantly decrease pixel noise and lower the statistical measurement uncertainties.

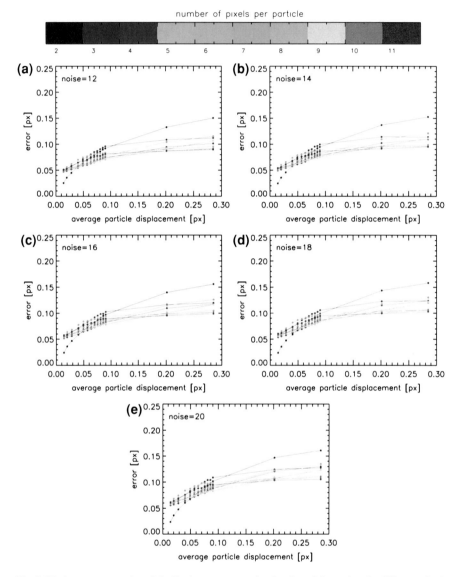

Fig. 9.17 Average error of particle displacements vs. simulated particle motion for different pixel-noise levels 12–20 in units of intensity. Number of pixels per particle are color-coded

9.3.3 Pixel-Noise from Intensity Fluctuations

Sometimes it is not possible to obtain a pixel-noise distribution from the background, e.g. if there was a cutoff intensity set during recording, or if the background is too dark, which often happens at very high frame rates (short exposure times). Nevertheless,

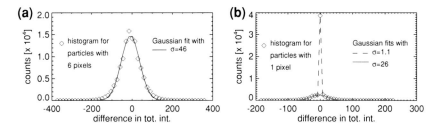

Fig. 9.18 Example of a histogram of $\Delta I\,(N_P)$ and the Gaussian fit for **a** $N_P = 6$ and **b** $N_P = 1$. In the *right plot*, the *solid line* is a fit without considering the central peak

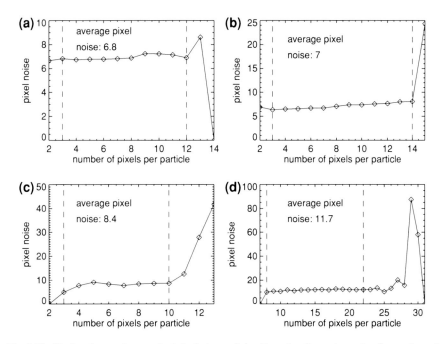

Fig. 9.19 Pixel noise as the standard deviations of the Gaussian fits to intensity fluctuation of particles consisting of N_P pixels, divided by $\sqrt{2N}$. For the averaged value of the pixel noise only values within the dashed lines were used. The data are from experiments rI (**a**) and rII (**b**) of the recrystallization experiments, and the two data sets used in the calculation of the coupling strength I (**c**) and II (**d**)

to obtain an estimate of the uncertainties, the intensity fluctuations of individual particle images from one frame to the next are considered (R. Sütterlin, private communication). At high frame rates the particles in an undisturbed crystal move only a very small distance from one frame to the next. Therefore, one can assume that the particle does not move into regions with different illumination during this time, nor does the particle change its height within the laser sheet.

Then the total intensity $I_{\text{tot},k}(t)$ of all pixels N_P assigned to a particle should not change from one frame to the next. Any fluctuations observed are assumed to be due to pixel noise.

The difference $\Delta I = I_{\text{tot},k}(t_2) - I_{\text{tot},k}(t_1)$ for particle k is calculated for consecutive frames as long as its number of pixels stays constant. For good statistics, the largest possible amount of frames and particles of a measurement should be considered. The distribution of the ΔI of all particles with the same number of pixels N is then fitted by a Gaussian function; examples are shown in Fig. 9.18. If a particle changed its number of pixels from one frame to the next, it was omitted. This procedure yields values for the pixel noise for different sizes of particles. The width of the fitted distributions, divided by $\sqrt{2N_P}$, is taken as the magnitude of the error. The factor $1/\sqrt{2N_P}$ comes from Gaussian error propagation. The errors for different particle sizes should be equal since they all come from the same pixel noise. All those were averaged where the number of points (the number of available particles with a certain size) was large enough to have a reliable statistical result. This then gives an estimate for the pixel noise in that peculiar experiment. Figure 9.19 shows the results for the experiments which were presented in this thesis.

Chapter 10
Appendix B: Bond Orientational Order

To get an idea of the influence of non-ideal crystal structures on the behavior of the bond correlation function $g_6(r)$, an artificial crystal with 4020 particles is generated and different transformations are applied to the particle coordinates to simulate density gradients, domain forming and random particle motion. Then the bond correlation analysis is performed on the simulated data and the shape of $g_6(r)$ is compared to bond correlation functions obtained in the experiments.

The lattice vectors $\mathbf{r}$ which point to the respective lattice sites of an ideal hexagonal lattice are given by:

$$\mathbf{r} = n_1 \mathbf{a} + n_2 \mathbf{b} \quad \text{with } n_1, n_2 \in \mathbb{Z} \tag{10.1}$$

and the elementary lattice vectors

$$\mathbf{a} = \Delta \begin{pmatrix} 1 \\ 0 \end{pmatrix}, \quad \mathbf{b} = \Delta \begin{pmatrix} 0.5 \\ \sin(\pi/3) \end{pmatrix} \tag{10.2}$$

Around each lattice site a unit cell is defined as the Wigner-Seitz cell, i.e. this cell is the smallest unit to choose so that still all space between the lattice sites is covered by adjacent cells.

The bond correlation function $g_6(r)$ gives information on the nature of the long range orientational order. Its shape indicates how uniform the bond angles are directed across the system when bonds separated by a distance r are compared with each other. $g_6(r)$ is defined as

$$g_6(r) = \left| \frac{1}{N_B} \sum_{l=1}^{N_B} \frac{1}{n(l)} \sum_{k=1}^{n(l)} \exp\{i6(\theta(\mathbf{r}_k) - \theta(\mathbf{r}_l))\} \right| \tag{10.3}$$

with N_B the total number of bonds in the crystal, $n(l)$ the number of bonds at the distance r from bond l and $\theta(\mathbf{r}_{k,l})$ being the respective angles of bonds at $\mathbf{r}_{k,l}$ to an arbitrary axis. Note that $g_6(r)$ is always 1 for a perfect hexagonal lattice by

C. A. Knapek, *Phase Transitions in Two-Dimensional Complex Plasmas*,
Springer Theses, DOI: 10.1007/978-3-642-19671-3_10,
© Springer-Verlag Berlin Heidelberg 2011

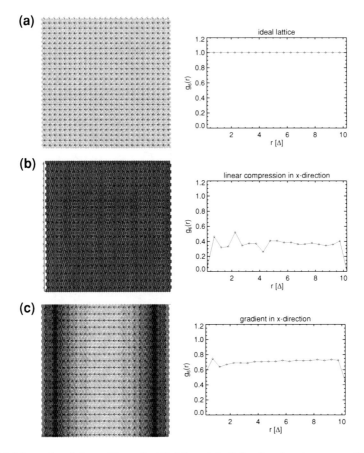

Fig. 10.1 Maps of particle positions of artificial crystals (*left column*) and corresponding $g_6(r)$ (*right column*). The unit cells are color coded with respect to their bond order parameter (*lighter colors* mean a higher value of Ψ_6). The *arrows* show in the direction of ϕ. **a** ideal lattice, **b** linear compression along x (*horizontal direction*), **c** compression with a gradient along the x-axis

definition. The bond order parameter Ψ_6 will be used to measure the local quality of the hexagonal structure. It is calculated from the complex quantity Ψ_6

$$\Psi_{6,k} = \frac{1}{n} \sum_{j=1}^{n} e^{6i\theta_{kj}} = \Psi_{6,k} e^{\{i\phi\}} \tag{10.4}$$

$$\phi = \arctan\{\Im(\Psi_{6,k})/\Re(\Psi_{6,k})\} \tag{10.5}$$

and is also 1 for the ideal hexagon.

10.1 Ideal Crystal

In the ideal crystal, $g_6(r) = 1$ for all r. All unit cells are perfect hexagons and their orientation is uniform. Figure 10.1a shows the map of the lattice (left image) with the cells encircled by white lines. The color corresponds to the bond order parameter Ψ_6, calculated for each cell (brighter colors are closer to ideal hexagons), and the arrows point into the direction of the argument ϕ of Ψ_6. In the right column of Fig. 10.1, the corresponding $g_6(r)$ is plotted.

10.2 Compression

The crystal was compressed in x-direction (horizontal axis in the figures) by $x = 0.6x_{\text{ideal}}$. The orientation of cells is still uniform in Fig. 10.1b, but the darker color shows a decrease in local order due to the distortion of the cells. Further $g_6(r)$ dropped to ≈ 0.4 but is still constant.

The results for a compression with a symmetric gradient $x = x_{\text{ideal}} \pm 0.1x_{\text{ideal}}^2$ in x-direction is presented in Fig. 10.1c. Due to the better hexagonal structure in the middle region the bond correlation function is constant at ≈ 0.7, which is higher than in the linear compression. The orientation of cells is uniform except in thin stripes along the outer vertical edges.

10.3 Rotation of Areas of the Crystal

The crystal is now divided into four equally sized parts and those parts are rotated with respect to each other by an angle $d\theta$, where r_{ideal}, θ_{ideal} are the polar coordinates of the particle in the ideal lattice:

$$\begin{pmatrix} x \\ y \end{pmatrix} = r_{\text{ideal}} \begin{pmatrix} \cos\left(\theta_{\text{ideal}} + d\theta\right) \\ \sin\left(\theta_{\text{ideal}} + d\theta\right) \end{pmatrix}$$

Figure 10.2a shows the case of a small angle rotation with angles $\leq 1°$. The effect is nearly non-existent, apart from a small decrease of $g_6(r)$ for all r, and the appearance of slightly less perfect unit cells in the domain boundaries.

At rotation angles of $15°$ (Fig. 10.2b) the domain boundaries contain defects symbolized by red (5-fold) and blue (7-fold) dots. The orientation of unit cells differs considerably from one domain to the next. The local order is still high, and does not seem to be affected much by the rotation, except at the domain boundaries. A linear decay of $g_6(r)$ is clearly visible in the right part of Fig. 10.2b.

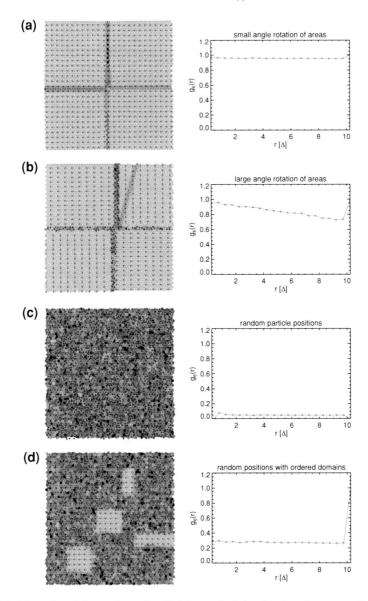

Fig. 10.2 Maps of particle positions of artificial crystals (*left column*) and corresponding $g_6(r)$ (*right column*). The unit cells are color coded with respect to their bond order parameter (lighter colors mean a higher value of Ψ_6). The arrows show in the direction of ϕ. Red and blue dots mark positions of 5-fold and 7-fold defects, respectively. **a** rotation of domains by small angles $\leq 1°$, **b** rotation of domains by large angles of $15°$, **c** random particle distribution, **d** random particle positions with a small number of ordered domains

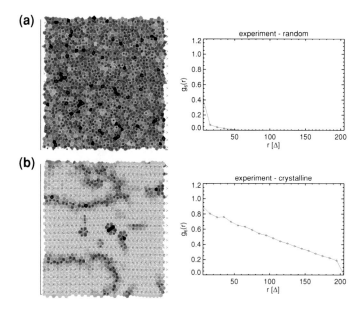

Fig. 10.3 Maps of particle positions of experimental data (*left column*) and corresponding $g_6(r)$ (*right column*). The unit cells are color coded with respect to their bond order parameter (lighter colors mean a higher value of Ψ_6). The arrows show in the direction of ϕ. *Red* and *blue dots* mark positions of 5-fold and 7-fold defects, respectively. **a** random particle distribution, liquid-like state, **b** crystalline state

10.4 Random Particle Positions

A small random component ξ is added to the particle coordinates, $\begin{pmatrix} x \\ y \end{pmatrix} = \begin{pmatrix} x_{\text{ideal}} \\ y_{\text{ideal}} \end{pmatrix} +$ ξ. In this particle distribution, $g_6(r)$ stays practically constant at less than 0.1 from the beginning. A large amount of defects exist, and there is no local order or preferred direction (Fig. 10.2c).

A small number of highly ordered domains were inserted into the randomly distributed system, shown in Fig. 10.2d. It rises the magnitude of $g_6(r)$ to ≈ 0.3, independent of r.

10.5 Experimental Bond Correlations

The results for the artificial crystals are compared to two images from the experiment (Fig. 10.3a, b). The first example is from a state with high disorder (Fig. 10.3a) and a large defect fraction. Here $g_6(r)$ drops to 0 fast within less than 3 interparticle distances. The second example Fig. 10.3b is taken from a much more crystalline state. $g_6(r)$ drops linear with r, as in the case of domain rotation by large angles (Fig. 10.2b). In the particle map, one can see the existence of domains with different orientation of unit cells, separated by chains of defects. Apparently the linear decay of $g_6(r)$ is connected to the domains with different lattice orientation.

Curriculum Vitae

Christina Ariane Knapek

Date of birth   26 August 1976
Place of birth   München, Germany
Nationality   German

Educational Qualifications

1986–1995   Secondary education
             Theresien-Gymnasium München (degree: Abitur)
1995–2004   Studies of general physics
             Technische Universität München (degree: Diplom)
             Diploma thesis: Experimental Investigation of Dynamical
             Properties and Ergodicity in Plasma Crystals
2004–2010   PhD student
             Thesis topic: Phase transitions in two-dimensional complex
             plasmas
             Thesis supervisor: Prof. Dr. Gregor Morfill
             Place: Max-Planck-Institut für extraterrestrische Physik, Garching
             and Ludwig-Maximilians-Universität, München, Germany

Working Experience

Since 2008   Research position in PlasmaLab (*Investigation of Complex Plasmas
             under Microgravity on the ISS—Development of new Plasma
             Chambers*), Max-Planck-Institut für Extraterrestrische Physik,
             Garching, Germany

C. A. Knapek, *Phase Transitions in Two-Dimensional Complex Plasmas*,   157
Springer Theses, DOI: 10.1007/978-3-642-19671-3,
© Springer-Verlag Berlin Heidelberg 2011

Publications in Refereed Journals

1. V. Nosenko, S. K. Zhdanov, A. V. Ivlev, C. A. Knapek and G. E. Morfill, 2D Melting of Plasma Crystals: Equilibrium and Nonequilibrium Regimes, Phys. Rev. Lett. **103 (1)**, 015001 (2009).
2. K. Rypdal, B. Kozelov, S. Ratynskaia, B. Klumov, C. Knapek and M. Rypdal, Scale-free vortex cascade emerging from random forcing in a strongly coupled system, NJP **10**, 093018 (2008).
3. G. E. Morfill, A. V. Ivlev, M. Rubin-Zuzic, C. A. Knapek, R. Pompl, T. Antonova and H. M. Thomas, Complex plasmas—new discoveries in strong coupling physics, Appl. Phys. B **89 (4)**, 527–534 (2007).
4. S. Ratynskaia, R. Kompaneets, A. V. Ivlev, C. Knapek, G. E. Morfill, Transport in strongly coupled two-dimensional complex plasmas: Role of the interaction potential, Phys. Plasmas **14 (1)**, 010702 (2007).
5. C. A. Knapek, A. V. Ivlev, B. A. Klumov, G. E. Morfill and D. Samsonov, Kinetic characterization of strongly coupled systems, Phys. Rev. Lett. **98 (1)**, 015001 (2007).
6. C. A. Knapek, D. Samsonov, S. K. Zhdanov, U. Konopka and G. E. Morfill, Recrystallization of a 2D plasma crystal, Phys. Rev. Lett. **98 (1)**, 015004 (2007).
7. S. Ratynskaia, K. Rypdal, C. Knapek, S. Khrapak, A. V. Milovanov, J. J. Rasmussen and G. E. Morfill, Superdiffusion and viscoelastic vortex flows in a two-dimensional complex plasma, Phys. Rev. Lett. **96 (10)**, 105010 (2006).
8. S. Ratynskaia, C. Knapek, K. Rypdal, S. Khrapak, G. E. Morfill, Statistics of particle transport in a two-dimensional dusty plasma cluster, Phys. Plasmas **12 (2)**, 022302 (2005).

Publications in Conference Proceedings

1. U. Konopka, M. Schwabe, C. Knapek M. Kretschmer, G. E. Morfill, Complex plasmas in strong magnetic field environments, 4th International Conference on Physics of Dusty Plasmas, New Vistas in Dusty Plasmas, AIP Conference Proceedings 799, 181–184 (2005).
2. C. Knapek, D. Samsonov, S. Zhdanov, U. Konopka, G. E. Morfill, Structural properties and melting of 2D-plasma crystals, 4th International Conference on Physics of Dusty Plasmas, New Vistas in Dusty Plasmas, AIP Conference Proceedings 799, 231–234 (2005).